Mathematical Marvels: Texts and Monographs in the Spirit of CR Rao

This series aims to publish books in all areas influenced by Dr. C.R. Rao, including statistics, linear algebra, econometrics and information geometry. It also includes content in other domains of mathematical sciences, applied statistics, financial mathematics, operations research and computer science with primary focus on mathematics and statistics. The series aims to carry on the tradition of C.R. Rao's innovative and influential works, which have impacted a variety of fields, and to provide top-notch textbooks and monographs in related areas.

George Osipenko

Symbolic Analysis of Dynamical Systems

Springer

George Osipenko
Department of Applied Mathematics
Lomonosov Moscow State University,
Sevastopol Branch
Moscow, Russia

ISSN 3005-1967 ISSN 3005-1975 (electronic)
Mathematical Marvels: Texts and Monographs in the Spirit of CR Rao
ISBN 978-981-95-0712-2 ISBN 978-981-95-0713-9 (eBook)
https://doi.org/10.1007/978-981-95-0713-9

Mathematics Subject Classification: 05C85, 37A05, 37D05, 34C37, 37B10, 37C20, 37C70, 39A33

This Springer imprint is published by the registered company Springer Nature Singapore Pte Ltd.
The registered company address is: 152 Beach Road, #21-01/04 Gateway East, Singapore 189721, Singapore

Preface

This book was written on the basis of a special course on Computer Simulation of Dynamic Systems. The lectures were devoted to methods for studying the global properties of dynamic systems and differ from well-known monographs in the original research method called the Symbolic Image of Dynamic System. The symbolic image is a directed graph that approximates a dynamic system by discretizing phase space. The dynamics produced by such a graph reflects the dynamics of the system. The symbolic image provides a unified basis for applying various techniques to dynamic systems whether they are continuous or discrete in time. By converting the system's flow into a graph, researchers can express their methods in terms of graph algorithms. Symbolic images are both an effective theoretical tool and the foundation for computer-oriented techniques for numerically studying non-local properties of systems. The monograph is a continuation of "Dynamical systems, graphs, and algorithms," *Lecture Notes in Mathematics*, Volume 1889 (2006). If the results have already been proven in that book, they are presented here without proof. Proofs of the remaining results can be found in full. We present current theoretical and applied results on symbolic image analysis for dynamic systems. This material fully reflects the current state of research in this area. A symbolic image is an effective tool for theory and the basis for numerical methods for investigating non-local system properties. From a computing perspective, this approach based on graphs offers significant benefits: once a symbolic image has been established, all future research can be focused on analyzing the graph. Lectures include numerous illustrative examples for comprehension. The book is aimed at researchers studying dynamic systems and their application in various scientific and technical fields, including engineering, physics, chemistry, biology, economy, and demography. Lectures will be of interest to students, graduate students, and teachers of natural science departments.

Moscow, Russia

George Osipenko
george.osipenko@mail.ru

Contents

About the Author

George Osipenko is a Professor in the Department of Applied Mathematics, Lomonosov Moscow State University, branch in Sevastopol, Russia, since 2015. Earlier, he taught at the Institute of Banking in Sevastopol, from 2010 to 2014. He taught at the Sevastopol Polytechnic Institute, from 2006 to 2009. He also taught at the St. Petersburg (Leningrad) Polytechnic Institute of Peter the Great, from 1975 to 2005. He completed his graduation and post-graduation in mathematics from the Faculty of Mathematics and Mechanics of the Leningrad University, Russia, respectively, in 1972 and 1975. He defended his doctoral dissertation and received a professor's degree, in 1990. He is awarded the N. N. Bogolyubov award, in 2009.

Chapter 1
Introduction

1.1 Dynamic Systems

A discrete dynamical system is generated by a mapping $f : M \to M$, where M is a bounded closed domain in R^d or a compact manifold. It is assumed that f is a homeomorphism, meaning that f is a continuous map that possesses a continuous inverse mapping f^{-1}. The state of the system is represented by a point $x \in M$, and the dynamics are determined by the iterative equation:

$$x_{n+1} = f(x_n), \tag{1.1}$$

where the subsequent state of the system x_{n+1} is uniquely determined by the previous state x_n through the mapping f. The Eq. (1.1) is referred to as the difference equation, and the integers $n \in Z$ are treated as discrete time. The evolution of the system's state is characterized by a sequence of points $\{x_n, x_{n+1} = f(x_n), \text{n} \in Z\}$, which is called the trajectory or orbit of the initial point $x = x_0$ and is denoted by $Tr(x)$. The set M is regarded as the phase space of the system. Theoretically, it is convenient to consider M as a compact Riemannian manifold.

The discrete dynamical system exhibits the group property:

$$f^{n+m}(x) = f^n\left(f^m(x)\right);\ n, m \in Z.$$

A continuous dynamical system is typically represented by an autonomous system of differential equations:

$$\frac{dx}{dt} = F(x). \tag{1.2}$$

G. Osipenko, *Symbolic Analysis of Dynamical Systems*, Mathematical Marvels: Texts and Monographs in the Spirit of CR Rao, https://doi.org/10.1007/978-981-95-0713-9_1

Let $\Phi(t, x)$ be a solution to the system (1.2) with initial condition $\Phi(0, x) = x$, and let the mapping $\Phi(t, x)$ be defined for $t \in R$ and $x \in M$. The evolution of the continuous system is given by the curve $\{x = \Phi(t, x_0), t \in R\}$, which is referred to as the trajectory of the point x_0. A continuous dynamical system also adheres to the group property:

$$\Phi(t + s, x) = \Phi(t, \Phi(s, x)); \; s, t \in R.$$

A continuous dynamical system generates a discrete system through the mapping $f(x) = \Phi(1, x)$, which serves as the shift operator along the trajectories per unit time of the system defined by (1.2). In this context, the dynamics of the resulting discrete system determines the dynamics of the continuous system represented by (1.2).

A periodic system of differential equations is characterized by the form:

$$\frac{dx}{dt} = F(t, x), \tag{1.3}$$

where $F(t, x)$ is a periodic mapping in t with period T. Let $\Phi(t, x)$ be a solution to the system (1.3) with initial condition $\Phi(0, x) = x$. The system (1.3) generates a discrete system through the mapping $f(x) = \Phi(T, x)$, which is the shift operator along trajectories over the period of the system (1.3). The mapping $\Phi(T, x)$ is referred to as the Poincaré mapping, named after Henri Poincaré, who introduced it and demonstrated that the dynamics of the resulting discrete system determine the dynamics of the system defined by (1.3).

Formulation of the Problem

Let the dynamic system be defined by the Eq. (1.1). Our goal is to determine the global properties of the system dynamics using computational methods without any prior knowledge about the system's characteristics. All information regarding the dynamics can be derived from the given mapping $f : M \to M$. To achieve this, we will develop and justify a computer-oriented technology for studying dynamics, based on the analysis of a symbolic image.

1.2 Coding

The modern theory and practice of dynamical systems pose the challenge of describing the global structure of trajectories and integral properties of the system dynamics. A critical tool for studying complex phenomena such as chaos and strange attractors is the trajectory encoding method. The name highlights the main idea: to describe dynamics by encoding trajectories as sequences of symbols.

As an example, consider the encoding of points in the Cantor set, which is generated from the initial segment [0, 1] as follows. The segment is divided into three equal parts, and the middle (open) interval is discarded. Each of the remaining segments is then divided into three parts again, and the middle part is discarded, continuing this process indefinitely. The outcome of endlessly repeating this procedure is the Cantor set. Each point in this set is obtained as the limit of nested segments, whose length tends to zero. We encode the sequence of nested segments with two symbols: 0 and 1. At each step, we choose one of the two segments: the left or the right. The choice of the left segment is denoted by 0, while the choice of the right segment is denoted by 1. For example, the sequence 0110... indicates that the left segment was selected in the first step, the right segment in the second and third steps, and the left segment again in the fourth step. Hence, an infinite sequence of zeros and ones encodes a series of nested segments, ultimately converging to a point in the Cantor set. Every point in the Cantor set corresponds uniquely to a sequence of zeros and ones. Notably, instead of using 0 and 1, we can employ any two symbols; for instance, we could use 'l' and 'r' to represent left and right. In this case, each point in the Cantor set would correspond to an infinite sequence (word) of two letters. This encoding's structure is independent of the specific symbols used; it is determined solely by their sequential alternation. For the set of all sequences composed of two symbols, we can introduce a distance that is generated by the distance in the original segment. For definitiveness, we assume our initial segment is [0, 1], and the coding is performed using the symbols 0 and 1. With each division, the number of segments doubles, and the length of each individual segment decreases by a factor of three. After N divisions, the length of each segment is given by $\frac{1}{3^N}$. Let ω and σ be two sequences of zeros and ones encoding the numbers α and β from the Cantor set. If the first N elements in these sequences are the same, but the (N + 1)-th elements differ, then the numbers α and β lie within the same segment of length $\frac{1}{3^N}$. However, during the next division, they end up in different segments, separated by a segment of length $\frac{1}{3^{N+1}}$. Therefore, the distance between α and β satisfies the inequalities:

$$\frac{1}{3^{N+1}} \le \rho(\alpha, \beta) \le \frac{1}{3^N}.$$

We can define the distance between the sequences ω and σ as either $1/(3^N)$ or$1/(3^{N+1})$. Similarly, we can define the distance for sequences composed of any number of symbols.

1.3 Dynamics on Graph

Let $S = \{a, b, c, ...\}$ be a finite set of symbols. Consider a set of sequences that are infinite in both directions:

$$\Sigma = \{x = \{\dots x_{-2}, x_{-1}, x_0, x_1, x_2, \dots\}\}$$

where each x_i is a symbol from S. We can introduce a distance in the sequence space Σ as previously described. The dynamics in the sequence space are determined by a shift mapping of the form:

$$y = \sigma(x), \text{ where } y_i = x_{i+1}$$

The mapping $\sigma : \Sigma \to \Sigma$ obviously has an inverse mapping σ^{-1}. Both mappings are continuous in the topology under consideration, making σ a homeomorphism.

At times, we consider the space Σ restricted to only include valid sequences, referred to as admissible sequences. For instance, let G be a directed graph, where each vertex v has outgoing and incoming edges. In this case, the symbol set $S = \{v\}$ corresponds to the set of vertices, and the space Σ consists of sequences of vertices $\{v_i, i \in Z\}$ that correspond to valid paths in the graph. The dynamics constructed in this manner form a topological Markov chain and are supported by a well-developed theory (see [2]). Consider a topological Markov chain with a fixed transition probability p_{ij} for each admissible transition i → j. In this scenario, we can represent the transitions with an $n \times n$ matrix:

$$P = \left(p_{ij}\right).$$

Additionally, transitions of the form i → i are permitted, assuming that the transition from state i to some state occurs with a probability of 1 per unit time. This leads to the condition that the sum of the elements in each row of the matrix P equals 1:

$$\sum_j p_{ij} = 1$$

The matrix P is referred to as the stochastic matrix, and the associated dynamics are called a stochastic Markov chain. Stochastic chains are widely utilized and possess a well-developed theory (see [3]).

1.4 Determinism and Limited Accuracy

Let the dynamics of the system be determined by the difference Eq. (1.1). In this case, the future state of the system is uniquely determined by the initial state x_0. This raises the question: how does the dynamics change when the initial value x_0 is altered? Lyapunov's theory of stability (see [4]) provides conditions for the continuous dependence of trajectories on the initial conditions.

At the end of the nineteenth century, Henri Poincaré, while studying the three-body problem, discovered bi-asymptotic trajectories that approach a certain periodic trajectory as $t \to \infty$ and as $t \to -\infty$. Such a double asymptotic trajectory is now referred to as a homoclinic trajectory. In the vicinity of a homoclinic trajectory, even an arbitrarily small change in the initial value x_0 can result in a trajectory that becomes periodic or returns to an arbitrarily small neighborhood of x_0. Therefore, even a minor change in x_0 can alter the topology of the trajectory; that is, the trajectory is dependent on the exact value of x_0. As a result, an imperceptible change in the initial data can lead to significant alterations in the trajectory.

In the 1960s, D. York introduced the term "chaos" to describe such phenomena. Stephen Smale demonstrated (see [5]) that near a homoclinic trajectory, there exists an invariant set Ω such that the dynamics on Ω are equivalent to the shift map $\sigma: \Sigma \to \Sigma$ in the space of binary sequences, where there exists a dense trajectory in Ω. Thus, an arbitrarily small perturbation to x_0 generates a dense trajectory whose points reside in every small neighborhood around each point in the set Ω. The dynamics along such a trajectory resemble a random walk along the invariant set Ω. It is important to note that both Poincaré dynamics and Smale dynamics are fully deterministic processes. Thus, chaotic behavior can manifest even within deterministic systems.

References

1. Alekseev, V.M.: Symbolic dynamics. 11th mathematical school. Kiev, 1976 (in Russian)
2. Lind, D., Marcus, B.: An introduction to symbolic dynamics and coding. Cambridge University Press, (1995)
3. Kemeney, J.G., Snell, J.L.: Finite Markov chains. Univ. Ser. Undergrad. Math.. Princeton, Van Nostrad (1960)
4. Lyapunov, A. M.: The general problem of the stability of motion. Comm. Mathematical Society of Kharkov, 1892 (in Russian). Engl. Transl. Int. J. Control., **55**(3), 531–773, (1992) Taylor and Francis Ltd.
5. Smale S.: Diffeomorphisms with many periodic points. Differ. Comb. Topol.. Princeton Univ. Press, (1965), 63–80

Chapter 2
Symbolic Image of Dynamic System

2.1 Definition

Let $f : M \to M$ be a homeomorphism of a compact manifold M, generating the discrete dynamical system given by

$$x_{n+1} = f(x_n). \tag{2.1}$$

We denote the distance on M by $\rho(x, y)$. In the following discussion, we will utilize the concept of a symbolic image of a dynamical system [1], which effectively merges trajectory coding [2, 3] with numerical methods [4, 5].

Let $C = \{M(1), ..., M(n)\}$ be a finite covering of a manifold M by closed subsets, see Fig. 2.1. The set $M(i)$ will be referred to the cell of the index i. Symbolic image of the dynamic system (2.1) for a covering C is a directed graph G with vertices $\{i\}$ corresponding to cells {M(i)}. Two vertices i and j are connected by a directed edge (an arc) $i \to j$ if and only if

$$f(M(i)) \cap M(j) \neq \emptyset.$$

Let us consider a geometric example of constructing a symbolic image. Let M be a closed domain on the plane, as depicted in Fig. 2.1. The covering C consists of 100 cells, numbered from the upper left corner from left to right and from top to bottom. Thus, the upper left cell has an index 1, the index of the upper right cell is 10, the bottom left cell's index is 91, and the bottom right cell's is 100. Consider cell $M(i)$, where $i = 63$. Let the image f(M(i)) intersect with cells with indices $l = 26, k = 27, j = 28, e = 36$, and $d = 37$. As a result, we have five edges emanating from i toward j, k, l, e, and d. By repeating this process for each cell, we obtain the symbolic image. If f(M(i)) intersects with cell M(i), a loop $i \to i$ is formed.

G. Osipenko, *Symbolic Analysis of Dynamical Systems*, Mathematical Marvels: Texts and Monographs in the Spirit of CR Rao, https://doi.org/10.1007/978-981-95-0713-9_2

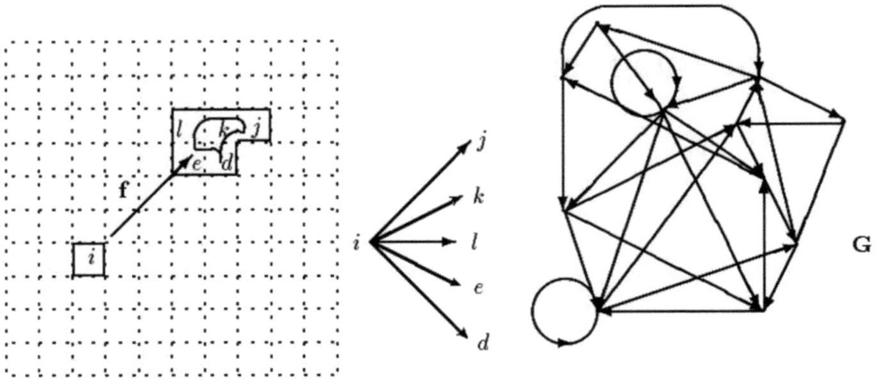

Fig. 2.1 Construction of a symbolic image

The directed edges of G represent the potential paths that trajectories can take as they evolve over time. Importantly, the presence of an edge does not imply that all points in $M(i)$ will eventually map to points in $M(j)$; it simply indicates that at least one point does so. This aspect is crucial when analyzing the system's structure, as it highlights potential points of overlap between different subsets of the manifold. The symbolic image allows us to encode the trajectories generated by the iterations of the function f. By associating sequences of points in M with symbolic sequences, we can analyze the system's behavior in a more tractable format. This encoding can reveal the underlying structure of the dynamics and facilitate the study of stability, periodic orbits, and chaotic behavior. Through this approach, we aim to obtain a deeper understanding of the qualitative and quantitative aspects of the dynamical system defined by f.

We do not impose any special restrictions on the covering C, but based on the theorem on the triangulation of a compact manifold, we can assume without loss of generality that the cells M(i) are polyhedra intersecting along boundary disks. In numerical calculations (see [6]), we consider M as a compact domain in R^d, where the cells M(i) are represented as cubes or parallelepipeds.

In the section on measure theory, we will refine our analysis by transitioning from the covering C to a more structured partition denoted by C^*. This partition is constructed by assigning each boundary disk to one of the adjoining. The resulting partition C^* is a measurable partition of the manifold M. By defining the measurable partition C^*, we can apply measure-theoretic methods to investigate various properties of the dynamical system, such as **Invariant Measures** and **Ergodic Properties.**

2.2 Simulation Parameters

Definition 2.1. A sequence of vertices $\{z_k\}$ in a graph G, extending infinitely in both directions, is called a path (or admissible path) if, for each vertex $k \in Z$, the graph contains a directed arc connecting the vertices $z_k \to z_{k+1}$.

We denote the vertex set of the graph by Ver. The symbolic image G can be considered as a multi-correspondence between the vertices $G : Ver \to Ver$, where the image $G(i)$ is the set of vertices that are the terminal points of arcs $i \to j$ or

$$G(i) = \{j : i \to j\}.$$

There exists a natural multi-valued mapping $h : M \to Ver$ that maps the manifold M onto the vertices $Ver : h(x) = \{i : x \in M(i)\}$. From the definition of a symbolic image, it follows that the diagram

$$\begin{array}{ccc} M & \xrightarrow{f} & M \\ \downarrow h & & \downarrow h \\ Ver & \xrightarrow{G} & Ver \end{array} \tag{2.2}$$

is commutative in the following sense:

$$h(f(x)) \subset G(h(x)). \tag{2.3}$$

Indeed, let $i \in h(x)$ and $j \in h(f(x))$. Then $M(j) \cap f(M(i)) \neq \emptyset$ and there is an arc $i \to j$, which means $j \in G(i)$ or $h(f(x)) \in G(h(x))$. This means $h(f(x)) \in G(h(x))$. We cannot guarantee that $h(f(x)) = G(h(x))$. However, the inclusion (2.3) is sufficient to state that the trajectories of f are transformed by the mapping h onto the paths of the symbolic image

$$h(Tr) = \omega = \left\{i_n : f^n(x) \in M(i_n)\right\}.$$

The trace ω can be interpreted as an encoding of the trajectory Tr.

Now we introduce some parameters of the symbolic image. Let

$$diamM(i) = max(\rho(x, y) : x, y \in M(i))$$

be the diameter of the cell $M(i)$, and let $d = diam(C)$ be the largest diameter among the cells in the covering C. The parameter d is referred to as the diameter of the covering.

Let q (referred to as the upper bound of the symbolic image) be the largest diameter of the images

$$q = max\{diam f(M(i))\}.$$

Since the mapping f has a modulus of continuity $\theta(\rho)$, the estimate $q \leq \theta(d)$ holds.

We define the number r as follows. By construction, the cells $M(i)$ are closed sets. If $M(k) \cap f(M(i)) = \emptyset$, then the distance

$$r_{ik} = \rho(f(M(i)), M(k)) = min\{\rho(x, y) : x \in f(M(i)), y \in M(k)\}$$

is positive. Since the number of pairs described is finite,

$$r = min\{r_{ik} : M(k) \cap f(M(i)) = \emptyset\}$$

is positive. Therefore, r represents the smallest distance between the images and the cells such that $M(k) \cap f(M(i)) = \emptyset$. The value r is called the lower bound of the symbolic image.

Proposition 2.1 [7] If a point x is in cell $M(j)$ and $\rho(x, f(M(i))) < r$, then the image $f(M(i))$ intersects cell $M(j)$.

2.3 Calculation of Symbolic Image

Petrenko, in his dissertation [8], examined five methods for numerically constructing a symbolic image. Here, we will focus on one of these methods. It is important to note that the numerical construction of a symbolic image involves approximation, as the image of a cell is computed approximately.

The following straightforward method of numerical construction yields good results. Let the covering C consist of polyhedra. In each cell $M(i)$, a fixed number of points $\{x_k, k = 1, 2, \cdots, N\}$ are uniformly distributed. These points are referred to as "scanning points." Typically, for smooth systems, $10 \leq \mathrm{N} \leq 100$ is adequate. The images $f(x_k)$ are then calculated. Furthermore, we assume that an arc $i \to j$ exists if there is a scanning point $x_k \in M(i_k)$ such that $f(x_k) \in M(j)$. Thus, each arc of the symbolic image is represented on the manifold by a pair of points $x_i \to f(x_i)$. By fixing the described pair for each arc $i \to j$, we project the symbolic image onto the manifold.

2.4 Coding of Pseudo-trajectories

A sequence of points $Tr = \{x_n\}$, extending infinitely in both directions, is considered a trajectory of system (2.1) if it satisfies conditions $f(x_n) = x_{n+1}$.

Definition 2.2 *A sequence of points* $\omega = \{x_n\}$, *also extending infinitely in both directions, is called an* ε*-trajectory or a pseudo-trajectory if it meets the following condition:*

$$\rho(f(x_n), x_{n+1}) < \varepsilon$$

for any n.

In practice, calculations are not performed with absolute accuracy. Consequently, as a result of these calculations, we obtain a pseudo-trajectory.

Example 2.1 Consider a mapping of the form

$$f(x, y) = \big(y, 0.05\big(1 - x^2\big)y - x\big).$$

and take a sequence of points (see Fig. 2.2).

$$x_1 = (2, 0), x_2 = (0, -2), x_3 = (-2, 0), x_4 = (0, 2), x_{k+4} = x_k.$$

Let us check that this sequence is a 4-periodic ε-trajectory for any $\varepsilon > 0.1$. Indeed, $f(2, 0) = (0, -2); f(0, -2) = (-2, -0.1); f(-2, 0) = (0, 2), f(0, 2) = (2, 0.1)$. Thus, $|x_{k+1} - f(x_k)| \leq 0.1$ for $k = 1, 2, 3, 4$.

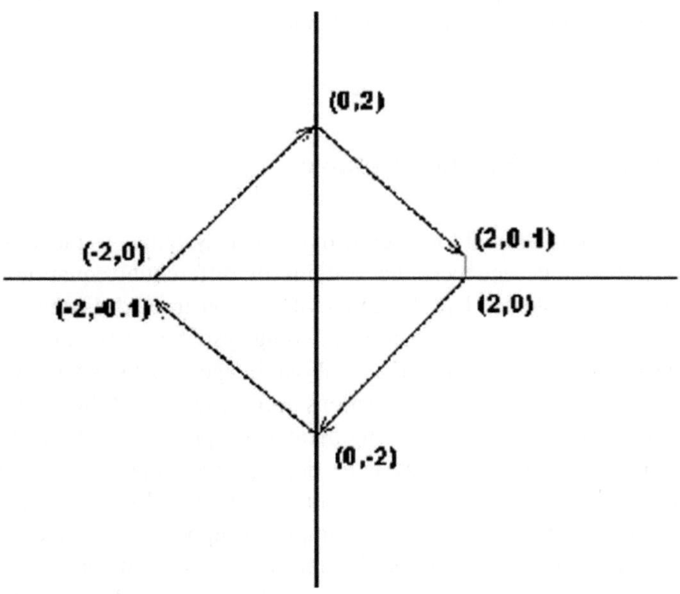

Fig. 2.2 Pseudo-trajectory of the mapping $f(x, y) = \big(y, 0.05\big(1 - x^2\big)y - x\big)$

There is a natural correspondence between admissible paths on the symbolic image and ε-trajectories of the homeomorphism f. Roughly speaking, an admissible path serves as the trace of an ε-trajectory, and vice versa.

Theorem 2.1 Weak shadowing property [7].

1. *If a sequence* $\{z_k\}$ *is a path on the symbolic image G and another sequence* $\{x_k\}$ *is such that satisfies specific conditions* $x_k \in M(z_k)$*, then sequence* $\{x_k\}$ *is an* ε*-trajectory for any given* $\varepsilon > d + q$*. In particular, if the sequence* $\left\{z_1, z_2, \cdots, z_p = z_0\right\}$ *is a p-periodic path, then the sequence* $\left\{x_1, x_2, \cdots, x_p = x_0\right\}$ *is a p-periodic* ε*-trajectory for any* $\varepsilon > d + q$*.*
2. *If a sequence* $\{z_k\}$ *is a path on the symbolic image G, then there exists a sequence* $\{x_k\}, x_k \in M(z_k)$ *that is an* ε*-trajectory for any given* $\varepsilon > d$*. In particular, if sequence* $\left\{z_1, z_2, \cdots, z_p = z_0\right\}$ *is a p-periodic path on the symbolic image, then there exists a sequence* $\left\{x_1, x_2, \cdots, x_p = x_0\right\}$ *that is a p-periodic* ε*-trajectory for any* $\varepsilon > d$*.*
3. *If a sequence* $\{x_k\}$ *is an* ε*-trajectory,* $\varepsilon < r$ *and* $x_k \in M(z_k)$*, then the sequence formed by that trajectory is an admissible path* $\{z_k\}$ *on the symbolic image. In particular, if the sequence* $\left\{x_1, x_2, \cdots, x_p = x_0\right\}$ *is a p-periodic* ε*-trajectory, then the sequence* $\{z_k\}$ *is an admissible p-periodic path on the symbolic image.*

According to the first and second parts of Theorem 2.1, each admissible path in the symbolic image can be interpreted as an encoding of pseudo-trajectories. Each trajectory $Tr = \{x_n = f^n(x_0)$ generates a path $\omega(Tr) = \{i_n\}$ such that $x_n \in M(i_n)$ an encoding or trace of the trajectory can be defined.

2.5 Application of Symbolic Image

The symbolic image is defined for discrete dynamical systems. As mentioned previously, both autonomous and periodic systems of differential equations generate discrete dynamical systems. The dynamics of the constructed discrete system, in turn, determine the dynamics of the corresponding system of differential equations. This relationship allows us to apply the symbolic image to the study of autonomous and periodic systems of differential equations according to the following scheme:

First, it is essential to construct a shift operator that progresses along the trajectories per unit of time for autonomous systems, or per period for periodic systems. Next, we construct a symbolic image for the resulting mapping that defines the discrete dynamical system. Article [6] presents examples of applying a symbolic image to mappings that may be discontinuous or may not have an inverse mapping. A particular example of using a symbolic image involves the localization of the chain-recurrent set of the Julia map $z \to z^2 + c, z \in C$. It is important to note that the Julia mapping generates a semidynamical system because it lacks an inverse mapping.

2.6 Transition Matrix

The directed graph is uniquely defined by its $n \times n$ (adjacency) transition matrix $\Pi = (\pi_{ij})$, where an entry π_{ij} is 1 if there is a directed edge between the vertices, and 0 otherwise. The i-th row of the matrix corresponds to the i-th vertex of the graph. Consequently, changing the numbering of the vertices alters the matrix of admissible transitions. Let

$$\Pi^2 = \left(\pi_{ij}^2\right)$$

be the square of the transition matrix, where the superscript 2 denotes an index (not a power). The sum $\sum_{k=1}^{n} \pi_{ik}\pi_{kj}$ of the elements provides the number of all admissible paths of length 2 from vertex i to vertex j. Similarly, it can be shown that the element π_{ij}^p of the degree of the matrix Π^p indicates the number of paths of a given length p from vertex i to vertex j. Notably, the trace of the matrix Π^p

$$tr\Pi^p = \sum_{i=1}^{n} \pi_{ii}^p$$

counts all p-periodic paths on the symbolic image.

Definition 2.3 *A vertex in the symbolic image is called recurrent if there exists a periodic path that passes through it. The set of recurrent vertices is denoted by RV. Two recurrent vertices i and j are regarded as equivalent if there exists a periodic path containing both i and j. The set of recurrent vertices is then partitioned into classes $\{H_k\}$ of equivalent recurrent vertices.*

In graph theory, these classes are known as strongly connected components. Algorithms for constructing strongly connected components based on a depth-first search of the graph are well-established [9, 10], with the most popular being the algorithm developed by Robert Tarjan [11, 12].

From a theoretical perspective, the matrix of permissible transitions serves as a useful tool for studying symbolic images. However, for the symbolic image to accurately model the system's dynamics, it is necessary to employ a sufficiently fine covering. This need often results in a large number of cells, and the matrix of admissible transitions may have many elements. It is advisable to focus on a manageable number of these elements. Typically, the number of non-empty intersections will be less than ten, leading to the i-th row of the matrix containing k ones and (n-k) zeros.

The matrix of admissible transitions is sensitive to the numbering of vertices or coverage cells; changing the numbering alters the form of the matrix. However, there exists a numbering that reflects the system's dynamics, allowing the transition matrix to assume a certain canonical form. We can introduce a quasi-order relation among the vertices of the symbolic image such that $i \longmapsto j$ if there is a path

$$i = i_0, i_1, i_2, \ldots, i_m = j.$$

A vertex i will be recurrent if and only if there is an admissible path linking it to itself. Furthermore, a pair of recurrent vertices will be considered equivalent if there exists such a periodic path connecting them.

Proposition 2.2 [2]. *The vertices of a symbolic image G can be renumbered so that*

- *the equivalent recurrent vertices are numbered with consecutive integers;*
- *-the new numbers i, j of other vertices are chosen so that $i < j$ if there is a path $i \longmapsto j$, but no path $j \longmapsto i$.*

In summary, the transition matrix can be structured such that

$$\Pi = \begin{pmatrix} \Pi_1 & * & * & \cdots * * \\ 0 & \ddots & * & \cdots * * \\ 0 & 0 & \Pi_k & \ddots \vdots \vdots \\ \vdots & \ddots & \ddots & \ddots \; * \; \vdots \\ \vdots & \ddots & \ddots & \ddots \; \ddots \; * \\ 0 & 0 & \cdots & \cdots \; 0 \; \Pi_s \end{pmatrix} \begin{matrix} H_1 \\ \vdots \\ H_k \\ \vdots \\ \vdots \\ H_s \end{matrix},$$

where each diagonal block corresponds either to an equivalence class of recurrent vertices or to a non-recurrent vertex represented by a zero. Below the diagonal blocks, only zeros exist. This type of transition matrix is referred to as upper triangular. The described renumbering of vertices is known as the canonical sorting of the graph.

2.7 Subdivision Procedure

Definition 2.4 *Let Q and G be abstract directed graphs. A mapping $s : Q \to G$ is called a graph mapping if it consistently transforms the vertices and edges of graph Q into the vertices and edges of graph G. In other words, if k and l are vertices in Q connected by an edge $k \to l$, and if $s(k) = i$ and $s(l) = j$ are their corresponding vertices in G, then the edge $i \to j$ must exist in G and $s(k \to l) = i \to j$. The converse must also be true: if and $s(k \to l) = i \to j$ exists in G, then $s(k) = i$ and $s(l) = j$.*

A mapping of directed graphs generates a corresponding mapping of admissible paths. Notably, a periodic path is transformed into another periodic path. It should be noted that, under such a transformation, the period may decrease. Recurrent vertices map to recurrent vertices, and equivalent recurrent vertices map to equivalent recurrent vertices. Hence, strong components are also transformed into strong components.

Let's apply the adaptive subdivision procedure to the coverings and construct a sequence of symbolic images. We will consider the main stage of the subdivision procedure. Let $C = \{M(i)\}$ be a covering of the manifold M, and let G be the symbolic image corresponding to this covering. Assume that a new covering NC is a subdivision of the covering C. This means that each cell $M(i)$ in the original covering is subdivided into smaller cells, denoted as $m(i, k)$, $k = 1, 2, ...,$ such that

$$\bigcup_k m(i, k) = M(i).$$

Let NG denote the symbolic image for the new covering $NC = \{m(i, k)\}$. The vertices of the new symbolic image are denoted by (ik). It is possible that some cells will not be subdivided, i.e., $m(i, 1) = M(i)$. This subdivision generates a natural mapping s from NG onto G, which transforms a vertex (i, k) in NG to the vertex i in G; that is, $s(i, k) = i$.

The intersections for small cells ensure that similar intersections hold for larger cells. Consequently, s transforms a directed edge $(i, k) \to (j, l)$ into the directed edge $i \to j$. Therefore, s maps the directed graph NG to the directed graph G. As a mapping of directed graphs, s takes every admissible path in NG to some admissible path in G. In particular, the image of a periodic path remains a periodic path, and the image of a recurrent vertex remains a recurrent vertex. Moreover, the image of the class NH of equivalent recurrent vertices in NG lies within the class H of equivalent recurrent vertices in G.

2.8 Sequence of Symbolic Images

Consider the sequence of coverings $\{C_t\}$ of the manifold M by cells obtained through successive subdivisions; that is, coverage cells C_{t+1} are obtained by dividing the cells of coverage C_t. Let us denote these coverage cells as $M(z_t)$, where z_t is the cell index. As the diameter of the covering tends to zero, we consider the sequence of symbolic images $\{G_t\}$ of the mapping f with respect to the coverings C_t. This gives us a commutative diagram

$$\begin{array}{ccccccc} V_1 & \overset{s_1}{\leftarrow} & V_2 & \overset{s_2}{\leftarrow} & V_3 & \overset{s_3}{\leftarrow} & \cdots \\ G_1 \downarrow & & G_2 \downarrow & & G_3 \downarrow & & \\ V_1 & \overset{s_1}{\leftarrow} & V_2 & \overset{s_2}{\leftarrow} & V_3 & \overset{s_3}{\leftarrow} & \cdots \end{array}, \tag{2.4}$$

where $s_t(z_{t+1}) = z_t$ if $M(z_{t+1}) \subset M(z_t)$.

Let's consider the limit case when each cell consists of a single point, meaning that the vertex set corresponds to the points in M equipped with the discrete topology. The arc set consists of pairs of the form $\{x \to f(x)\}$. In this case, we can regard the above-defined symbolic image as coinciding with the original mapping $f : M \to M$.

Let G be the symbolic image for the coverage $C = \{M(i)\}$, and let V be the set of vertices. We have defined a multi-valued mapping $h : M \to V$ that associates a point x with all vertices i such that $x \in M(i)$. The following diagram is commutative:

$$\begin{array}{ccc} V & \stackrel{h}{\leftarrow} & M \\ G\downarrow & & f\downarrow; \\ V & \stackrel{h}{\leftarrow} & M \end{array}$$

Taking this into account, we can supplement the diagram (2.4) accordingly.

$$\begin{array}{ccccccc} V_1 & \stackrel{s_1}{\leftarrow} & V_2 & \stackrel{s_2}{\leftarrow} & V_3 & \stackrel{s_3}{\leftarrow} \dots \stackrel{h}{\leftarrow} & M \\ G_1\downarrow & & G_2\downarrow & & G_3\downarrow & & f\downarrow \\ V_1 & \stackrel{s_1}{\leftarrow} & V_2 & \stackrel{s_2}{\leftarrow} & V_3 & \stackrel{s_3}{\leftarrow} \dots \stackrel{h}{\leftarrow} & M \end{array}$$

We omit the subscript t in the notation for s_t when confusion cannot arise. Each mapping s forms a directed graph that maps an admissible path to another admissible path. By the definition of a subdivision, if $s(i) = j$, then the cell $M(i)$ lies within the subdivision of cell $M(j)$.

Let P_t be the space of admissible paths on the symbolic image G_t. The mapping s: $G_{t+1} \to G_t$ of directed graphs defines the mapping of the paths

$$s : P_{t+1} \to P_t, s(P_{t+1}) \subset P_t.$$

Let $T = \{x_k = f^k(x_0), k \in Z\}$ be a trajectory.

According to Theorem 2.1, each trajectory $Tr = \{x_k\}$ generates an admissible path of the form $\chi^t = \{z^t(k), x_k \in M(z_t(k))\}$ on each symbolic image G_t. Furthermore, these paths are interconnected by mapping s:

$$s(\chi^{t+1}) = \chi^t.$$

The constructed path χ^t can be considered an encoding of the trajectory Tr with respect to the covering C_t.

According to Theorem 2.1, each path on a symbolic image generates a pseudo-trajectory. If a dynamical system possesses the pseudo-trajectory tracking property (see [13]), then the tracking corresponds to the actual trajectory (assuming the covering diameter is sufficiently small). Thus, to effectively encode trajectories, it is sufficient to construct a symbolic image of a covering with a sufficiently small diameter, negating the need for an infinite sequence of subdivisions.

References

1. Osipenko, G. S.: On the symbolic image of a dynamical system. Bound. Value Problems. Interuniv. Collect. Sci. Work., Perm State Tech. Univ., Perm, 101–105 (1983) (in Russian)
2. Alekseev, V.M.: Symbolic dynamics. 11th mathematical school. Kiev, (in Russian) (1976)
3. Lind, D., Marcus, B.: An introduction to symbolic dynamics and coding. Cambridge University Press, (1995)
4. Hsu, C.S.: Cell-to-Cell Mapping. Springer-Verlag, NY (1987)
5. Nusse, H., Yorke, J.: Dynamics: numerical exploration. Springer New-York Berlin (1997)
6. Avrutin, V., Levi, P., Schanz, V., Fundinger, D., Osipenko G.: Investigation of dynamical systems using symbolic images: efficient implementation and applications. International Journal of Bifurcations and Chaos **16**(12), 3451–3496 (2006)
7. Osipenko, G.: Dynamical systems, graphs, and algorithms. Lectures Notes in Mathematics, 1889, Springer, Berlin, (2007)
8. Petrenko, E.I.: Computer research of dynamic systems based on the symbolic image method. Ph.D. dissertation, St. Petersburg University, (2009)
9. Cormen, T.H., Leiserson, C.E., Rivest, R.L., Stein, C.: Introduction to algorithms. MIT Press, Cambridge (2002)
10. Siek, J.G., Lie-Quan Lee, Lumsdaine, A.: C++ Boost graph library. Addison-Wesley, Boston—San Francisco—New York, (2001)
11. Tarjan, R.: Depth-first search and linear graph algorithms. SIAM J. Comput. **1**, 146–160 (1972)
12. Tarjan, R.: Algorithm design. Commun. ACM **30**(3), 204–212 (1987)
13. Pilyugin, Yu.S.: Shadowing in dynamical system. Lect. Notes Math., 1706, Springer, Berlin, (1999)

Chapter 3
Shadowing of Trajectories

3.1 Consistent Paths

Let $\{C_t\}$ be a sequence of closed coverings, each of which is obtained through the subdivision procedure, and let $\{G_t\}$ be the corresponding sequence of symbolic images. If we have a path ω^t on each symbolic image G_t, we then obtain a sequence of paths $\{\omega^t \in P_t\}$, where P_t is the set of admissible paths on G_t. The paths ω^t for different t are generally not related to each other.

The mapping h(x) associates a point x with the index of the cell in which it resides. According to Theorem 2.1, any trajectory $T(x_0) = \{x_k = f^k(x_0), k \in Z\}$ defines an admissible path of the form

$$\omega^t = h(T(x_0)) = \{z^t(k) : x_k \in M(z^t(k))\},$$

where $z^t(k) \in h(x_k)$. The described path serves as the coding of the trajectory $T(x_0)$ with respect to the covering. The subdivision generates a natural mapping s from C_{t+1} to C_t, which was outlined in the previous section that means

$$\omega^t = s(\omega^{t+1}).$$

This observation leads us to the following definition:

Definition 3.1 *Let $\{C_t\}$ be a sequence of closed coverings, each obtained by subdividing the previous covering. A sequence of paths $\{\omega^t \in P_t\}$ is called consistent if the equality*

$$\omega^t = s(\omega^{t+1}) \tag{3.1}$$

holds for each t.

In what follows, we will need the concept of a recurrent trajectory. The definition of a recurrent trajectory varies among authors. For instance, Definition 3.2.2 of a

G. Osipenko, *Symbolic Analysis of Dynamical Systems*, Mathematical Marvels: Texts and Monographs in the Spirit of CR Rao, https://doi.org/10.1007/978-981-95-0713-9_3

recurrent trajectory in [1] aligns with the definition of a Poisson stable trajectory found in [2]. Here, we will adopt the definition of a recurrent trajectory in the sense of Birkhoff.

Definition 3.2 *A trajectory T is termed recurrent if for each $\varepsilon > 0$, there exists a positive integer $p > 0$ such that any interval of length p of this trajectory approximates the entire trajectory T with accuracy ε.*

A closed invariant set Ω is called minimal if it contains no non-trivial closed invariant subsets.

Theorem 3.1 Strong shadowing property. *Let $\{C_t\}$ be a sequence of closed coverings obtained* via *the subdivision procedure, with their diameters tending to zero. Suppose that on each G_t a path $\omega^t = \{i_k^t, k \in Z\}$ is defined such that the sequence of paths ω^t is consistent. Then the following results are valid:*

1. *There exists a unique trajectory $T = \{x_k : f(x_k) = x_{k+1}\}$ such that $x_k \in M(i_k^t)$ for any t.*
2. *If points $x_t(k)$ lie within the cells $M(i_k^t)$, then for each k, there exists a limit*

$$\lim_{t\to\infty} x_k^t = x_k$$

and the pseudo-trajectories $T_t = \{x_k^t \in M(i_k^t), k \in \mathbb{Z}\}$ converge to the trajectory T uniformly.

3. *The union of the cells of the path*

$$Q_t = \left\{ \bigcup_k M(i_k^t) : \{i_k^t\} = \omega^t \right\}$$

has the Following Properties:

3.1. *The sets Q_t are nested:*

$$Q_t \supset Q_{t+1},$$

3.2. *The limit of the compact sets Q_t is the closure of the trajectory T:*

$$\lim_{t\to\infty} Q_t = \bigcap_t Q_t = \overline{T}.$$

4. *If each path ω^t is periodic, then the shadowed trajectory T is recurrent, and $\lim_{t\to\infty} Q_t$ constitutes a minimal invariant set.*

Proof of the Statement 1. Let $\omega^t = \{i_k^t, k \in Z\}$ be the *path*. We fix k and consider the sequence of cells $\{M(i_k^t), t = 1, 2, ...\}$. Since $s(i_k^{t+1}) = i_k^t$, the cell $M(i_k^{t+1})$ belongs to the subdivision of the cell $M(i_k^t)$. Hence, we have the inclusions:

$$M(i_k^1) \supset M(i_k^2) \supset \cdots \supset M(i_k^t) \supset M(i_k^{t+1}) \supset . \tag{3.2}$$

Since the cells are closed and their diameters tend to zero together with d_t, there exists a unique point x_k in the intersection of all these cells

$$x_k = \lim_{t\to\infty} M\left(i_k^t\right) = \bigcap_t M\left(i_k^t\right).$$

Since the path $\omega^t = \left\{i_k^t, k \in Z\right\}$ is admissible, we have

$$f\left(M\left(i_k^t\right)\right) \cap M\left(i_{k+1}^t\right) \neq \emptyset$$

for any t. From (3.2) and $d_t \to 0$, it follows that the closed sets $\left\{f\left(M\left(i_k^t\right)\right), t = 1, 2, ...\right\}$ are nested, and their diameters q_t tend to zero. Hence, there exists a unique limit point

$$x_{k+1}^* = \lim_{t\to\infty} f\left(M\left(i_k^t\right)\right).$$

A similar analysis shows that the sequence of closed sets

$$\left\{f\left(M\left(i_k^t\right)\right) \cap M\left(i_{k+1}^t\right), t = 1, 2, ...\right\}$$

has a limit point

$$\lim_{t\to\infty} f\left(M\left(i_k^t\right)\right) \cap M\left(i_{k+1}^t\right) = x_{k+1}^{**}.$$

Since

$$x_{k+1}^{**} \in f\left(M\left(i_k^t\right)\right) \cap M\left(i_{k+1}^t\right) \subset M\left(i_{k+1}^t\right)$$

for any t, it follows that x_{k+1}^{**} lies in $M\left(i_{k+1}^t\right)$ as well.

Hence, we conclude that

$$\bigcap_t M\left(i_{k+1}^t\right) = x_{k+1}.$$

Hence, $x_{k+1}^{**} = x_{k+1}$. Because

$$x_{k+1}^{**} \in f\left(M\left(i_k^t\right)\right) \cap M\left(i_{k+1}^t\right) \subset f\left(M\left(i_k^t\right)\right)$$

for any t. We see that x_{k+1}^{**} lies in the intersection of all the cells

$$\bigcap_t M\left(i_{k+1}^t\right) = x_{k+1},$$

that is, $x_{k+1}^{**} = x_{k+1}$. Next,

$$x_{k+1}^{**} \in f\left(M\left(i_k^t\right)\right) \cap M\left(i_{k+1}^t\right) \subset f\left(M\left(i_k^t\right)\right)$$

for any t, and hence the point x_{k+1}^{**} lies in

$$\cap_t f\left(M\left(i_k^t\right)\right) = x_{k+1}^*,$$

that is, $f(x_k) = x_{k+1}$. Thus, Statement 1 has been proven.

Proof of the Statement 2. In the proof of Statement 1, we constructed the points

$$x_k = \lim_{t\to\infty} M\left(i_k^t\right) = \cap_t M\left(i_k^t\right)$$

and established that $f(x_k) = x_{k+1}$. Now, let's assume that the points x_k^t lie in the cells $M\left(i_k^t\right)$. Fixing k, we show that each sequence $\left\{x_k^t : t \in N\right\}$ converges to x_k. By our earlier reasoning, we have $x_k^t \in M\left(i_k^t\right)$ such that $x_k \in M\left(i_k^t\right)$. Hence,

$$\rho\left(x_k, x_k^t\right) \le diamM\left(i_k^t\right) \le d_t \to 0$$

and $\lim_{t\to\infty} x_k^t = x_k$. Since the distance $\rho\left(x_k, x_k^t\right) \le d_t$ is uniformly bounded over the sequence, the pseudo-trajectories $T_t = \left\{x_k^t, k \in Z\right\}$ converge uniformly to the trajectory T. Thus, Statement 2 is proven.

Proof of the Statement 3. Each covering C_t consists of a finite number of cells. The set

$$Q_t = \left\{\cup_k M\left(i_k^t\right) : \left\{i_k^t\right\} = \omega^t\right\}$$

is the union of a finite number of closed cells, therefore Q_t is compact. For a consistent sequence of paths ω^t, we have the equality $s\left(i_k^{t+1}\right) = i_k^t$, which gives the inclusion

$$M\left(i_k^{t+1}\right) \subset M\left(i_k^t\right).$$

Consequently, the sets Q_t are nested. Since every compact set Q_t contains the trajectory T and its closure, the intersection also contains the closure $\overline{T}$.

To show that the reverse inclusion $\cap_t Q_t \subset \overline{T}$ holds, let $x \in \cap_t Q_t$. Then x belongs to $\left\{\cup_k M\left(i_k^t\right) : \left\{i_k^t\right\} = \omega^t\right\}$ for any t. Consequently, x lies in some cell $M\left(i_k^t\right)$. Each cell $M\left(i_k^t\right)$ contains a point on the trajectory T according to Statement 1. There exists a point y_t on the trajectory T such that $\rho(x, y_t) \le d_t \to 0$ as $t \to \infty$. Thus, the sequence of points y_t on the trajectory converges to x. The third assertion has been proven.

Proof of the Statement 4. Assume each path $\omega^t = \left\{i_k^t, k \in Z\right\}$ is periodic with a period p_t, meaning $i_{k+p_t}^t = i_k^t$ for any $k \in Z$. Consider the path ω^1 on the symbolic image. According to the proof of the first statement of the theorem, the constructed trajectory $T = \{x_k : f(x_k) = x_k\}$ is contained in the union of cells:

$$U_1 = \left\{ \bigcup_k M\left(i_k^1\right) : i_k^1 \in \omega^1 \right\}.$$

By hypothesis, if ω^1 is a periodic path of the period p_1, then U_1 is the union of a finite number of cells $M\left(i_k^1\right)$, $k = 1, 2, \cdots, p_1$. Let d_1 be the diameter of the covering C_1. The ball B_k^1 with radius d_1 centered at point $x_k \in M\left(i_k^1\right)$ contains the cell $M\left(i_k^1\right)$. Thus, the union of balls

$$\bigcup_k B_k^1, \quad k = 1, 2, \cdots, p_1$$

contains the trajectory T. Consequently, the d_1-neighborhood of any interval of the trajectory T of length p_1 contains the entire trajectory T.

Let d_m be the diameter of the covering C_m and ω^m be the periodic path of period p_m on the symbolic image G_m. The similar reasoning for the covering C_m shows that the union of balls B_k^m of radius d_m centered at points $x_k \in M\left(i_k^m\right)$, $k = 1, 2, \cdots, p_m$ also contains the trajectory T. Therefore, the d_m-neighborhood of any interval of length p_m of the trajectory T contains the entire trajectory T. The trajectory T is recurrent since, for any positive ε, there exists $d_m < \varepsilon$ and a period p_m such that the d_m-neighborhood of any interval of the trajectory T of length p_m contains the entire trajectory T. It follows from Statement 3 that $\lim_{n\to\infty} Q_t$ is the closure of the trajectory. According to Birkhoff's theorem (see [2]), the closure of a recurrent trajectory is a compact minimal set. Thus, the theorem is proven.

Comment. According to Theorem 3.1, for a sequence of paths ω^t described in the condition of the theorem, there exists a unique trajectory T. However, the converse is not necessarily true; a given trajectory can generate more than one sequence ω^t of the described form. For instance, a trajectory that passes through a boundary point of the cells can generate multiple sequences of valid encodings.

3.2 Tikhonov Topology

Consider a directed graph G with vertices $V = \{i, j, \cdots\}$. Let $\omega = (i_k, k \in Z)$ be a path on G, and let $P = \{\omega\}$ be the space of paths. We can introduce a distance on P as follows, for example:

$$\rho(\omega_1, \omega_2) = \sum_{-\infty}^{\infty} \lambda^{|r|} \delta(i_r, j_r),$$

where $\omega_1 = (i_k, k \in \mathbb{Z})$, $\omega_2 = (j_k, k \in \mathbb{Z})$; $\delta(i, j) = 1$ if $i \neq j$, otherwise $\delta(i, j) = 0$; $0 < \lambda < 1$.

The space P is compact in this topology (see [3]). Additionally, this topology can also be generated by a different distance,

$$\rho(\omega_1, \omega_2) = \begin{cases} 2^{-k}, k = max\{\mathrm{m} : i_r = \mathrm{j_r} \ \text{for} \ |\mathrm{r}| \leq \ \mathrm{m}\}, \\ 0 \qquad\qquad\qquad\qquad\qquad npu \ \omega_1 = \omega_2 \end{cases}$$

which may result in a topology that reflects the convergence behavior of sequences of paths.

This means that the sequences of paths ω_1 and ω_2 coincide for $|\mathrm{r}| \leq \mathrm{m}$, but $i_{m+1} \neq i_{m+1}$ or $i_{-(m+1)} \neq i_{-(m+1)}$. The topological properties of an abstract sequence space equipped with the Tikhonov topology are discussed in [4].

Let $\{C_t, t \in N\}$ be a sequence of coverings of the manifold M by cells obtained through the subdivision procedure, and let their diameters d_t tend to zero. The space of paths on G_t equipped with the Tikhonov topology will be denoted by (P_t). The mapping $s : G_{t+1} \to G_t$ of directed graphs defines the projection in path space:

$$s : P_{t+1} \to P_t.$$

This mapping is continuous in the Tikhonov topology, and the image $s(P_{t+1})$ is compact in P_t.

Consider two subdivisions C_t and C_{t+m}. By construction, C_{t+m} is a subdivision of C_t. This gives rise to a mapping $s : G_{t+m} \to G_t$ of directed graphs and a corresponding mapping of path spaces $s : P_{t+m} \to P_t$. Suppose that on each symbolic image G_t, a path ω_t is given. Then, for each G_t, we obtain a sequence of paths $\omega_t^m = s(\omega_{t+m}), m = 1, 2, \cdots$, where each path ω_t^m is a projection of the path ω_{t+m} through the mapping $s : P_{t+1} \to P_t$.

Definition 3.3 *We say that the sequence ω_t weakly converges if ω_t^m converges in the Tikhonov topology as $m \to \infty$ for each fixed index t.*

Theorem 3.2 *Let $\{C_t, t \in N\}$ be a sequence of coverings of the manifold M obtained through the subdivision procedure, with their diameters d_t tending to zero, and assume that on each G_t there is a path $\omega_t = \{i_k^t, k \in Z\}$. If the sequence ω_t weakly converges and $x_k^t \in M(i_k^t)$, then the $\lim_{t\to\infty} x_k^t = x_k$ exists for each k, and the sequence $\{x_k, k \in Z\}$ is a trajectory of the mapping f.*

Proof. The space P_t is compact in the Tikhonov topology. Each path $\omega_t^m \subset P_t$ is a projection of the path ω_{t+m} through the mapping $s : P_{t+m} \to P_t$. According to the conditions of the theorem, the sequence $\{\omega_t^m, m \in \mathbb{N}\}$ converges as $m \to \infty$. We set

$$\lim_{m\to\infty} \omega_t^m = \omega_t^*.$$

Let's show that

$$s(\omega_{t+1}^*) = \omega_t^*.$$

In fact, $\omega_t^m = s(\omega_{t+m})$ and $\omega_{t+1}^m = s(\omega_{t+1+m})$. Since s is continuous (in the Tikhonov topology), we have

$$\lim_{m\to\infty} \omega_t^m = \lim_{m\to\infty} s(\omega_{t+m}) = \lim_{r\to\infty} s(\omega_{t+1+r}) = \lim_{r\to\infty} s\big(\omega_{t+1}^r\big) = s\Big(\lim_{r\to\infty} \omega_{t+1}^r\Big).$$

Therefore, the sequence ω_t^* is consistent. Let the path ω_t^* have the form $\{i_k^{*t}, k \in \mathbb{Z}\}$, and let x_k^{*t} be a point in the cell $M\left(i_k^{*t}\right)$. According to Assertion 2 of Theorem 3.1, for each k, the limit

$$\lim_{t\to\infty} x_k^{*t} = x_k$$

exists, and the sequence $\{x_k, k \in Z\}$ is a trajectory.

We show that $\lim_{t\to\infty} x_k^t = x_k$. According to the conditions of the theorem, each point x_k^t lies in the cell $M\left(i_k^t\right)$, where $\omega_t = \left\{i_k^t, k \in Z\right\}$. Fixing k and $\varepsilon > 0$, we can find τ such that the diameter of the covering d_τ becomes smaller than ε. For this τ, we find a number $m(l)$ such that the sequences ω_t^* and ω_t^m are sufficiently close in the Tikhonov topology so that

$$i_k^{*\tau} = s\big(i_k^{\tau+m}\big), \ |k| < l$$

for all $m > m(l)$. Thus, the equality

$$s\big(\omega_{t+1}^*\big) = \omega_t^*$$

implies that

$$s\big(\omega_{t+m}^*\big) = \omega_t^*.$$

Therefore, for the projections of the cell indices, we have

$$s\big(i_k^{*\tau+m}\big) = i_k^{*\tau} = s\big(i_k^{\tau+m}\big),$$

where s maps the directed graph $G_{\tau+m}$ onto G_τ.

This means that the points $x_k^{*\tau+m}$ and $x_k^{\tau+m}$ lie in the same cell $M\left(i_k^{*\tau}\right)$ with a diameter smaller than ε.

Therefore, for all $t > \tau + m$

$$\rho\big(x_k^t, x_k^{*t}\big) < \varepsilon, \ |k| < l.$$

Since ε is arbitrary, the equality

$$\lim_{t\to\infty} x_k^{*t} = x_k$$

confirms that

$$\lim_{t\to\infty} x_k^t = x_k$$

and the theorem is proven.

Sequence Space

Let Λ be the set of doubly infinite sequences $\xi = \{x_k, k \in Z\}$ of points from the manifold M. We consider x_k as the k-th coordinate of the sequence ξ. We introduce the distance r in the space Λ as follows. Let $\xi_1 = \{x_k, k \in Z\}$, $\xi_2 = \{y_k, k \in Z\}$ be two sequences and $\rho(x, y)$ be the distance on M. We define the distance r on Λ as

$$r(\xi_1, \xi_2) = \sum_{-\infty}^{\infty} \frac{\rho(x_k, y_k)}{2^{|k|}}.$$

This distance r is analogous to the distance between paths on a graph, generating the Tikhonov topology in the space of paths. We will refer to the above topology on the space of Λ as the Tikhonov topology.

When proving Statement 2 of Theorem 3.1, the sequence of paths $\{x_k^t\}$ converges to the sequence $\{x_k\}$ uniformly in k because $\rho(x_k^t, x_k) \le d_t$. In the proof of Theorem 3.2, the sequence $\{x_k^t\}$ converges to the trajectory $\{x_k\}$ in the Tikhonov topology of the space Λ because the paths ω_τ^* and ω_τ^m are close in this topology.

By the definition of the distance r, we have

$$\rho(x_k, y_k) < 2^{|k|} r(\xi_1, \xi_2).$$

Therefore, if ξ_1 converges to ξ_2 in the Tikhonov topology of the space Λ, then x_k converges to y_k on the manifold M; that is, the sequence ξ_1 converges coordinatewise to the trajectory ξ_2. The converse is also true.

Proposition 3.1 *Given a compact manifold M, if the sequences converge coordinatewise, then they also converge in the Tikhonov topology.*

Proof Consider sequences $\xi_t = \{x_k^t, k \in Z\}$, $\xi = \{y_k, k \in Z\}$ and $\xi = \{y_k, k \in \mathbb{Z}\}$. Let $\rho(x_k, y_k) \to 0$ as $t \to \infty$ for each k. The diameter of the compact manifold $d^* = max\{\rho(x, y); x, y \in M\}$ is a finite number. Therefore, $\rho(x_k^t, y_k) \le d^*$ for each k:

$$r(\xi_t, \xi) = \sum_{-\infty}^{\infty} \frac{\rho(x_k^t, y_k)}{2^{|k|}} = \sum_{-m}^{m} \frac{\rho(x_k^t, y_k)}{2^{|k|}} + \sum_{-\infty}^{-m-1} \frac{\rho(x_k^t, y_k)}{2^{|k|}} + \sum_{m+1}^{\infty} \frac{\rho(x_k^t, y_k)}{2^{|k|}} \le$$

$$\sum_{-m}^{m} \frac{\rho(x_k^t, y_k)}{2^{|k|}} + \sum_{-\infty}^{-m-1} \frac{d^*}{2^{|k|}} + \sum_{m+1}^{\infty} \frac{d^*}{2^{|k|}} =$$

$$\sum_{-m}^{m} \frac{\rho(x_k^t, y_k)}{2^{|k|}} + \frac{d^*}{2^{m-1}},$$

where m is a positive integer. Let us show that $r(\xi_t, \xi) \to 0$. Take any $\theta > 0$ and, for $\theta/2$, we find m such that

$$\frac{d^*}{2^{m-1}} < \frac{\theta}{2}.$$

For the found m, we define $t^* > 0$ so that the finite sum

$$\sum_{-m}^{m} \frac{\rho\left(x_k^t, y_k\right)}{2^{|k|}} < \frac{\theta}{2}.$$

For $t \geq t^*$. This is possible since m is fixed and every $\rho\left(x_k^t, y_k\right) \to 0$. In this case

$$r(\xi_t, \xi) < \theta.$$

Since θ is any positive number, $r(\xi_t, \xi) \to 0$ for $t \to \infty$. Hence, the proof is complete.

Corollary *The set of trajectories $Tr \subset \Lambda$ is a compact set in the Tikhonov topology.*

Proposition 3.2 If ξ_t *is a sequence of* ε_t*-trajectories* and it converges to zero, then there exists a subsequence ξ_{t_n} that converges to a trajectory in the Tikhonov topology.

Proof Let x_0^t be the zeroth element of the sequence ξ_t. Since the points $\left\{x_0^t, t \in N\right\}$ lie within a compact manifold, there exists a subsequence x_0^τ indexed by $\tau = t_m$ that converges to some point x^*. We aim to show that the trajectory passing through the point x^* is the one we require.

Consider the ε-trajectory $\xi = \{x_k, k \in Z\}$. We estimate the distance $\rho\left(x_k, f^k(x_0)\right)$ in terms of ε and the modulus of continuity θ of the mapping f:

$$\rho(f(x), f(y)) \leq \theta(\rho(x, y)),$$

where $\theta(\rho)$ can be assumed to be a monotonic function. For any $k > 0$, we have

$$\rho\left(x_k, f^k(x_0)\right) \leq \rho(x_k, f(x_{k-1})) +$$

$$\rho\left(f(x_{k-1}), f^2(x_{k-2})\right) + \rho\left(f^2(x_{k-2}), f^3(x_{k-3})\right) + \cdots \rho\left(f^{k-1}(x_1), f^k(x_0)\right).$$

We can estimate each term:

$$\rho(x_k, f(x_{k-1})) \leq \varepsilon,$$

$$\rho\left(f(x_{k-1}), f^2(x_{k-2})\right) \leq \theta(\rho(x_{k-1}, f(x_{k-2}))) \leq \theta(\varepsilon),$$

$$\rho\big(f^2(x_{k-2})\big), f^3(x_{k-3})) \le \theta(\rho\big(f(x_{k-2}), f^2(x_{k-3})\big) \le \theta^2(\varepsilon).$$

Here, θ^2 is the superposition of the function θ with itself. A similar analysis leads to the conclusion that

$$\rho\big(f^{k-1}(x_1)\big), f^k(x_0)) \le \theta^{k-1}(\varepsilon).$$

Ultimately, the continuity of f implies that the sequence ξ_τ converges to the trajectory $T(x^*)$ in the specified coordinates. From Proposition 3.1, it follows that ξ_τ converges to $T(x^*)$ in the Tikhonov topology. The proof is complete.

Theorem 3.3 Let $\{C_t, t \in N\}$ be a sequence of coverings of the manifold M by cells, obtained through successive subdivisions with diameters $d_t \to 0$, and suppose that a path $\omega_t = \left\{i_k^t, k \in Z\right\}$ is defined on each covering G_t. Then there exists a subsequence ω_τ, $\tau = t_m \to \infty$ such that, for any $x_k^\tau \in M\left(i_k^\tau\right)$, the limit

$$\lim_{\tau \to \infty} x_k^\tau = x_k$$

exists, and the sequence $\{x_k, k \in Z\}$ is a trajectory of the mapping f.

Proof Consider the symbolic image G_1 and the path space P_1, on G_1. We project all paths $\omega_t, t > 1$ onto P_1 via the mapping s. This creates a sequence of paths ω_t^1 in P_1. Since P_1 is a compact set, there exists a convergent subsequence ω_m^1, $m = t_r \to \infty$.

Next, we consider the symbolic image G_2 and the space of paths P_2, projecting all selected paths ω_m onto P_2. As before, we construct another convergent subsequence. By repeating this process, we construct a sequence ω_*^t, which forms a subsequence of the foregoing structure.

If we take the diagonal sequence ω_τ^τ, we arrive at a sequence of paths that converge in the Tikhonov topology. Now, the conclusion of this theorem follows directly from Theorem 3.2.

Theorem 3.4 Let U be a neighborhood of the set of trajectories Tr in the Tikhonov topology. Then there exists $d_0 > 0$ such that, for any symbolic image G constructed for a covering C with diameter $d \le d_0$, every trace $\chi(\omega)$ of any admissible path ω on G lies within the neighborhood U.

Proof Assume, for the sake of contradiction, that there exists a neighborhood U of the set of trajectories Tr for which the conclusion of the theorem does not hold. This assumption implies that for any $d > 0$, there exists a covering C with diameter d such that the symbolic image $G(C)$ contains an admissible path $\omega = \{i_k\}$ for which the sequence $\{x_k, x_k \in M(i_k)\}$ lies outside U.

According to Theorem 2.1, the sequence $\{x_k, x_k \in M(i_k)\}$ is an ε-trajectory of the mapping f for any $\varepsilon > q + d$, where q represents the maximal diameter of the cell images $f(M(i))$. We can take $\varepsilon = \theta(d) + d$ to be defined in terms of the modulus of continuity $\theta(d)$. Let C_t be a sequence of coverings with the subdivision diameters tending to zero, while fulfilling the aforementioned requirements. As a result, we obtain a sequence ε_t-trajectories $\xi_t = \{x_k^t\}$ with $\varepsilon_t \to 0$ lying outside the neighborhood U.

By Proposition 3.2, there exists a subsequence ω_{t_m} that converges to a trajectory T in the Tikhonov topology. However, this implies that T does not lie in the neighborhood U, leading to a contradiction since T should be in U. This contradiction completes the proof of Theorem 3.4.

3.3 Coding of Trajectories

Consider the symbolic image G of the mapping f. The set of admissible paths P on G is endowed with the Tikhonov topology, and Theorem 2.1 describes the connection between paths on G and pseudo-trajectories of the mapping f. Any trajectory $T = \{x_k, k \in Z\}$ has its trace on G in the form of the path $\omega = \{i_k : x_k \in M(i_k)\}$, which can naturally be referred to as the encoding of the trajectory. The multivalued mapping $h(x) = \{i : x \in M(i)\}$ defines the encoding of any trajectory. Let us denote the set of paths encoding trajectories as Cod. Our next goal is to describe the embedding $Cod \subset P$.

Let $\{C_t\}$ be a sequence of coverings of the manifold M by cells, obtained through successive subdivisions, and let their diameters d_t tend to zero. These coverings generate a sequence of symbolic images G_t and path sets P_t. The set of encodings Cod_t lies within P_t. The subdivision of cells defines a mapping $s : G_{t+1} \to G_t$ of directed graphs and a corresponding path mapping $s : P_{t+1} \to P_t$. The projection $s(P_{t+1})$ is compact in P_t.

Since the covering C_{t+m} is a subdivision of the covering C_t for any $m > 0$, there exists a mapping $s : G_{t+m} \to G_t$ of directed graphs and a projection $s : P_{t+m} \to P_t$ of path spaces. This results in a sequence of projections $P_t^m = s(P_{t+m})$ for $m > 0$ in the space of paths.

Theorem 3.5 *Let $\{C_t\}$ be a sequence of coverings of the manifold M by cells obtained through successive subdivisions, and let their diameters d tend to zero. Then:*

1. The projections $\{P_t^m\}$ form a sequence of nested compact sets, satisfying

$$P_t \supset P_t^1 \supset P_t^2 \supset \cdots \supset P_t^m \supset P_t^{m+1} \cdots ,$$

for all t.

2. *The set of encodings Cod_t is a compact set in P_t and*

$$Cod\_t = \lim_{m\to\infty} P_t^m = \bigcap_m P_t^m ,$$

3. *The s-image of the set of trajectory encodings coincides with the trajectory encodings*

$$s(Cod_{t+1}) = Cod_t.$$

Proof

1. The mapping $s : G_{t+1} \to G_t$ of directed graphs defines the path mapping $s : P_{t+1} \to P_t$, which is continuous in the Tikhonov topology. The image $s(P_{t+1}) = P_t^1$ is a compact set in P_t. The mapping: $P_{t+m} \to P_t$ is a composition of the mappings defined above. Consequently, all projections $P_t^m = s(P_{t+m}), m \in N$ are compact sets, and since $s(P_{t+1}) \subset P_t$, they represent nested compact sets $\{P_t^m, m \in N\}$.
2. For each t, we have inclusions $Cod_t \subset P_t$ and $s(P_{t+1}) \subset P_t$. Therefore, $Cod_t \subset P_t^m$ for any $m > 0$, and

$$Cod_t \subset \bigcap_m P_t^m.$$

   To prove the reverse inclusion, let $\omega \in \cap_m P_t^m$, then there exists a sequence $\omega_m \in P_{t+m}$ such that $s(\omega_m) = \omega$. Because the sequence ω_m is consistent, by Theorem 3.1, there exists a trajectory $T = \{x_k\}$, which is the trace of each path ω_m and ω. In other words, ω_m and ω are the encodings of the trajectory $T = \{x_k\}$, which demonstrates the inclusion $\omega \in Cod_t$.
3. To find the image of the trajectory encodings, we can assert that

$$s(Cod_{t+1}) = s\left(\bigcap_m P_{t+1}^m\right) = \bigcap_m s(P_{t+m}) = \bigcap_m P_t^m = Cod_t.$$

Thus, we have the desired equality. The theorem is proven.

3.4 Periodic Pseudo-trajectories

Let an ε-trajectory $\{x_n\}$ be periodic, meaning that there exists a period k such that $x_{n+k} = x_n$ for all n. The points x_n on this trajectory are called ε-periodic. In practice, the exact trajectory of the system is rarely known; instead, we generally work with ε-trajectories for sufficiently small positive values of ε.

Definition 3.4 *A point x is called chain-recurrent if x is ε-periodic for any $\varepsilon > 0$, that is, there is a periodic ε-trajectory passing through the point x.*

A chain-recurrent set consists of all chain-recurrent points and is denoted by CR. The chain-recurrent set CR is invariant and closed, containing all types of return trajectories: periodic, almost-periodic, non-wandering, and homoclinic. It is worth noting that if a chain-recurrent point is not periodic, there exists an arbitrarily small perturbation of the mapping f in the C^0-topology, for which this point becomes periodic [5]. Therefore, we can state that chain-recurrent points generate periodic trajectories under C^0-perturbations. Thus, in computational calculations, chain-recurrent points resemble periodic points of sufficiently large period.

Definition 3.5 *Two chain-recurrent points are called equivalent if they can be connected by a periodic* ε*-trajectory for any* $\varepsilon > 0$.

The chain-recurrent set is divided into equivalence classes $\{\Omega_i\}$*, which we will refer to as the components of the chain-recurrent set.*

It is important to note that there can be components that are not isolated from other components. For example, the equilibrium state $y = 0$ of the differential equation $y' = y^3 sin\left(1/y\right)$ is not an isolated equilibrium point. Let f be the shift along the trajectory of this equation per unit time. Each equilibrium state of the equation generates a fixed point of the diffeomorphism f, which is a component of the chain-recurrent set. The fixed point $y = 0$ is a component that is not isolated from other components.

The set of all ε-periodic points is denoted by $Q(\varepsilon)$.

Proposition 3.3 *The sets of* ε*-periodic points have the following properties:*

1. $Q(\varepsilon)$ *are open sets for* $\varepsilon > 0$,
2. If $\varepsilon_2 < \varepsilon_1$, *then*

$$Q(\varepsilon_2) \subset Q(\varepsilon_1). \tag{3.3}$$

Proof

1. Let $x \in Q(\varepsilon)$. By definition, there exists a periodic ε-trajectory $x_1, x_2, \cdots, x_p = x_0$ trajectory through x $= x_1$. We find $r_0 = max\{\rho(f(x_n), x_{n+1}), n = 1, 2, \cdots, p\}$ such that $r_0 < \varepsilon$. Since the manifold M is compact and the mapping f is continuous, there exists a modulus of continuity

$$\theta : \rho(f(x), f(y)) \leq \theta(\rho(x, y)).$$

Let $\delta > 0$ be a number such that $max\{\delta, \theta(\delta)\} < \varepsilon_0$.To show that all points y in the δ-neighborhood of the point x also lie in $Q(\varepsilon)$, we fix y and construct a periodic ε-trajectory passing through y. To do this, we replace the point x $= x_1$ with y in the sequence $x_1, x_2, \cdots, x_p = x_0$ and show that the new sequence is still an ε-trajectory. Indeed, if $\rho(x, y) < \delta$, then $\rho(f(x), f(y)) < \varepsilon_0$ and

$$\rho(x_2, f(y)) \leq \rho(x_2, f(x)) + \rho(f(x), f(y)) < r_0 + \varepsilon_0 = \varepsilon$$

In addition,

$$\rho(x_2, f(y)) \leq \rho(x_2, f(x)) + \rho(f(x), f(y)) < r_0 + \varepsilon_0 = \varepsilon$$

For $n = 2, \cdots, p-1$, the inequalities $\rho(f(x_n), x_{n+1}) < \varepsilon$ are preserved, since points $x_n, n = 2, \cdots, p$ do not change.

2. Since each ε_2-trajectory is a ε_1-trajectory for $\varepsilon_2 < \varepsilon_1$, we have

$$Q(\varepsilon_2) \subset Q(\varepsilon_1).$$

Each chain-recurrent point is ε-periodic for any $\varepsilon > 0$. Therefore, the chain-recurrent set CR can be obtained as the limit:

$$CR = \lim_{\varepsilon \to 0} Q(\varepsilon) = \bigcap_{\varepsilon > 0} Q(\varepsilon).$$

Thus, the family of open sets $\{Q(\varepsilon), \varepsilon > 0\}$ are nested, and the limit $\lim_{\varepsilon \to 0} Q(\varepsilon)$ is the chain-recurrent set. This means that $\{Q(\varepsilon), \varepsilon > 0\}$ forms a fundamental system of neighborhoods of the chain-recurrent set CR.

The Limit Set of a Sequence of Periodic Pseudo-trajectories.

Let $\{\chi_n\}$ be a sequence of periodic ε_n-trajectories. Let us fix a point x_n on each pseudo-trajectory χ_n. The sequence $\{x_n\}$ lies within a compact set in the manifold. Consequently, there exists a convergent subsequence $\{x_{n_k}\}$ that converges to some point x^*. This point x^* is referred to as a limit point, and the set of all limit points is called the limit set of the sequence $\{\chi_n\}$.

Proposition 3.4 *Let $\{\chi_n\}$ be a sequence of periodic ε_n-trajectories, $\varepsilon_n \to 0$. Then the limit set of the sequence $\{\chi_n\}$ consists of chain-recurrent points.*

Proof For any n, let x_n be a point on the pseudo-trajectory χ_n. Since the manifold is compact, there exists a convergent subsequence $\{x_{n_k}\}$ such that $x_{n_k} \to x^*$, implying that x^* is a limit point. We claim that x^* is a chain-recurrent point. To demonstrate this, we construct a sequence of periodic pseudo-trajectories $\{\omega_k\}$ by replacing the point x_{n_k} in the pseudo-trajectory χ_{n_k} with the limit point x^*. We will show that $\{\omega_k\}$ is an ε_k^*-trajectory with $\varepsilon_k^* \le \varepsilon_{n_k} + \gamma_k$, where $\gamma_k = max\{\theta(\delta_k), \delta_k\}$, δ_k is the distance between x_{n_k} and x^*, and $\theta(\rho)$ is the modulus of continuity for the mapping f. Indeed, by construction, we have

$$\rho\big(f\left(x_{n_k-1}\right), x_{n_k}\big) < \varepsilon_{n_k}, \; \rho\big(x_{n_k}, x^*\big) = \delta_k, \text{ and } \rho\big(f\left(x_{n_k-1}\right), x^*\big) < \varepsilon_{n_k} + \delta_k.$$

Moreover, since x_{n_k} converges to x^* and ε_{n_k} is arbitrary, we have $\rho\big(f\left(x_{n_k}\right), x_{n_k+1}\big) < \varepsilon_{n_k}$, $\rho\big(f\left(x_{n_k}\right), f(x^*)\big) < \theta(\delta_k)$ and $\rho\big(f(x^*), x_{n_k+1}\big) < \varepsilon_{n_k} + \theta(\delta_k)$. Since ε_{n_k} and δ_k both converge to 0, we have $\varepsilon_k^* \to 0$.

Thus, the sequences of periodic ε_k^*-trajectories $\{\omega_k\}$ are such that each $\{\omega_k\}$ passes through x^* and $\varepsilon_k^* \to 0$. Therefore, x^* is established as a chain-recurrent point. The proof is complete.

Note that when constructing the limit point, we transition from the sequence $\{\chi_n\}$ to the subsequence $x_{n_k} \in \chi_{n_k}$. This raises the question of the convergence of the original sequence $\{\chi_n\}$. However, there exist sequences of periodic ε_n-trajectories $\{\chi_n\}$ such that the overall sequence does not converge. To construct such an example, it suffices to consider two sequences of periodic pseudo-trajectories converging to different components of a chain-recurrent set. If the sequence is formed by interleaving these two sequences, then any resulting sequence $\{\chi_n\}$ fails to converge, since the limit must belong to different components of the chain-recurrent set.

Theorem 3.6 *Let $\{\chi_n\}$ be a sequence of periodic ε_n-trajectories, $\varepsilon_n \to 0$, and let x_n be a point on the pseudo-trajectory χ_n such that the limit point $x^* = \lim\limits_{n\to\infty} x_n$ lies within some component Ω of the chain-recurrent set. Then the sequence $\{\chi_n\}$ converges uniformly to Ω; that is,*

$$\rho(\chi_n, \Omega) = max_i\{\rho(x_i, \Omega), x_i \in \chi_n\} \to 0,$$

as $n \to \infty$.

Proof Let us assume the contrary, that there exists a sequence of points $y_n \in \chi_n$ such that $\rho(y_n, \Omega) > r > 0$. Since y_n lies within a compact set, there is a convergent subsequence $y_{n_k} \to y^*$. By Proposition 3.4, the limit point of this subsequence is a chain-recurrent point; however, it lies in a different component Ω^* since $\rho(y^*, \Omega) \geq r > 0$. Similarly, as shown in the proof of Proposition 3.4, we can construct a sequence of periodic ε_n-trajectories, $\varepsilon_n \to 0$, passing through the points x^* and y^*. This implies that the points x^* and y^* must lie in the same component of the chain-recurrent set, leading to a contradiction. Thus, our assumption is false, and the proof is complete.

Let $\{\chi_n\}$ be a sequence of periodic ε_n-trajectories, $\varepsilon_n \to 0$, and suppose the sequence of elements $x_0^n \in \chi_n$ converges to x^*. By Proposition 3.2, it follows that the sequence $\{\chi_n\}$ converges to the trajectory $T(x^*)$ of point x^* in the Tikhonov topology.

Definition 3.6 *A sequence $\chi_n = \left\{x_k^n, k \in Z\right\}$ of periodic ε-trajectories converges uniformly to the trajectory $T = \{y_k, k \in Z\}$ if $sup_k(\rho\left(x_k^n, y_k\right)$ converges to zero as n tends to infinity.*

Theorem 3.7 *If a sequence $\{\chi_n\}$ of periodic ε_n-trajectories, $\varepsilon_n \to 0$, converges uniformly to the trajectory T, then this trajectory is recurrent. The limit set of the sequence $\{\chi_n\}$ coincides with the closure of the trajectory T and forms a minimal set.*

Proof The pseudo-trajectory $\chi_n = \left\{x_k^n, k \in Z\right\}$ is periodic with period p_n. Let us fix n and set $r_n = sup_k\rho\left(x_k^n, y_k\right)$, $r_n \to 0$ as $n \to \infty$. Let $B(x, r)$ be a ball of radius r with center at point x. Since the pseudo-trajectory $\chi_n = \left\{x_k^n, k \in Z\right\}$ is p_n-periodic, the union of p_n balls

$$U_n = \left\{\bigcup_k B\left(x_k^n, r_n\right), x_k^n \in \chi_n\right\}$$

contains the entire trajectory $T = \{y_k, k \in Z\}$.

We will show that the trajectory T is recurrent. We fix $\varepsilon > 0$, find $r_n < {}^{\varepsilon}/_{2}$ which determines the number n corresponding to the periodicity p_n. From the above construction, it follows that the $2r_n$-neighborhood of any segment of the trajectory T of length p_n contains the entire trajectory T. Consequently, T is a recurrent trajectory. Since $r_n \to 0$ as $n \to \infty$, we see that the limit set of the sequence χ_n coincides with the closure of the trajectory T. The Birkhoff theorem (see [2]) states that the closure of a recurrent trajectory is a compact minimal set. Thus, theorem is proven.

3.5 Localization of Chain-Recurrent Set

Let G be a symbolic image with respect to a covering C and let d be the diameter of the covering. The multivalued mapping $h : M \to V$ associates with a point x all the vertices i such that x lies within the corresponding cell of the covering.

Proposition 3.5 *The mapping h has the following properties:*

1. *If x is a chain-recurrent point and $i \in h(x)$, then i is a recurrent vertex.*
2. *If x_1 and x_2 lie in the same component of the chain-recurrent set Ω, and $i_1 \in h(x_1)$ and $i_2 \in h(x_2)$ are the corresponding vertices, then i_1 and i_2 are equivalent recurrent vertices.*
3. *If Ω is a component of a chain-recurrent set, then there exists a unique class $H(\Omega)$ of equivalent recurrent vertices such that $h(\Omega) \subset H(\Omega)$.*

Proof

1. Let x be a chain-recurrent point, meaning that there exists a periodic ε-trajectory $\{x_1, x_2, \cdots, x_k = x_0\}$ passing through it $x = x_0$, $i \in h(x), x \in M(i)$. Choose $\varepsilon < r$ such that it is less than the lower bound of the symbolic image. Consider a periodic sequence of vertices $\{i_1, i_2, \cdots, i_k = i_0\}$, where $x_p \in M\left(i_p\right)$ for each p. According to Theorem 2.1, the periodic sequence $\{i_1, i_2, \cdots, i_k = i_0$ is an admissible periodic path on the symbolic image, it follows that $i = i_0$ is a recurrent vertex.
2. The second assertion can be proven similarly. If both x_1 and x_2 are chain-recurrent points in the same component Ω, then the periodic trajectories passing through them connect $h(x_1)$ *and* $h(x_2)$ establishing them as equivalent recurrent vertices.
3. Assertion 3 is a consequence of assertion 2. If Ω is a component of a chain-recurrent set, it follows from Proposition 3.5 that the image under the mapping h, denoted $h(\Omega)$, lies in some class of equivalent recurrent vertices, which we will denote as $H(\Omega)$. The component Ω lies in $\{\cup M(i) : i \in H(\Omega)\}$ as well. It is important to note that the class $H(\Omega)$ is uniquely determined by Ω. Thus, proposition is proven.

   Let us denote by $P(d)$ the union of the cells $M(i)$ for which the vertex i is recurrent, that is,

$$P(d) = \{\cup M(i) : \text{i is recurrent vertex}\} \tag{3.4}$$

Theorem 3.8

1. *Let* ε_1 *and* ε_2 *be such that* $0 < \varepsilon_1 < r < q + d < \varepsilon_2$,
 where r is the lower bound, q is the upper bound and d is the diameter of the covering for the symbolic image G. Then:

$$Q(\varepsilon_1) \subset P(d) \subset Q(\varepsilon_2), \tag{3.5}$$

 where $Q(\varepsilon)$ *is the set of* ε*-periodic points.*
2. *The set* $P(d)$ *is a closed neighborhood of the chain-recurrent set CR*

$$CR \subset P(d). \tag{3.6}$$

3. *For any neighborhood V of the chain-recurrent set CR, There exists* $d > 0$ *such that*

$$CR \subset P(d) \subset V.$$

4. *The chain-recurrent set CR coincides with the intersection of the sets* $P(d)$ *for all positive* d*:*

$$CR = \bigcap_{d>0} P(d). \tag{3.7}$$

Proof

1. If the point x lies in $Q(\varepsilon_1)$, then there exists a periodic ε_1-trajectory $\{x_1, ..., x_p = x_0\}$ passing through $x = x_1$.Consider the periodic sequence of cells $\{M(z_i) : i = 1, ..., p\}$ such that $x_i \in M(z_i)$. Since $\varepsilon_1 < r$, according to Theorem 2.1, the periodic sequence $\{z_i\}$ is an admissible periodic path passing through z_1. Therefore, the vertex z_1 is recurrent, and the cell $M(z_1)$ lies in $P(d)$.

Conversely, consider a point a point x lying in $P(d)$. There exists a cell $M(z)$ such that $x \in M(z)$ and the vertex z is recurrent. This implies that there exists a periodic path $\{z_1, z_2, ..., z_p\}$ that passes through $z_1 = z$. We can construct a periodic sequence of points $\{x_1, ..., x_p\}$ such that $x_i \in M(z_i)$. According to Theorem 2.1, the sequence $\{x_i\}$ constitutes a periodic ε_2-trajectory for any $\varepsilon_2 > q + d$. Thus, the point $x = x_1$ lies in $Q(\varepsilon_2)$.

2. From the inclusions $CR \subset Q(\varepsilon_1) \subset$ P(d) established in (1), we have that P(d) is a closed neighborhood of the chain-recurrent set CR.
3. Let V be a neighborhood of the set *CR*. Since the mapping f is uniformly continuous on the compact set M, the largest diameter q of the images of the cells tends to 0 as the covering diameter $d \to 0$. Thus, we can find d such that $\varepsilon_2 = {}^3/_2(q(d) + d) \leq \varepsilon^*$. Consequently, there exists ε^* such that the inclusion

$$P(d) \subset Q(\varepsilon_2) \subset Q(\varepsilon^*) \subset V$$

holds.

4. The family of open sets $\{Q(\varepsilon), \varepsilon > 0\}$ forms a fundamental system of neighborhoods of the chain-recurrent set CR. Therefore, we can write

$$CR = \bigcap_{\varepsilon>0} Q(\varepsilon).$$

From the inclusions (3.5), we get

$$CR \subset \bigcap_{d>0} P(d) \subset \bigcap_{\varepsilon_2} Q(\varepsilon_2) = CR,$$

where $\varepsilon_2 = {}^3/_2(q(d) + d)$ and $\varepsilon_2 \to 0$ as $d \to 0$. The proof is complete.

Assuming that $C = C_0$ is the original covering and $C_1, C_2, \cdots$ are obtained by the described subdivision process, the sets $P_0, P_1, P_2, \cdots$ are constructed according to (3.4).

Theorem 3.9 *The sequence of sets $P_0, P_1, P_2, \cdots$ has the following properties:*

(1) The sets P_k *are nested*

$$P_0 \supset P_1 \supset P_2 \supset \ldots \supset CR.$$

(2) If the covering diameters tend to zero, then

$$\lim_{k\to\infty} P_k = \bigcap_k P_k = CR,$$

where CR is the chain-recurrent set.

Proof

1. Let $C = \{M(i)\}$ be a covering of M and G be a symbolic image for a covering C. Assume that the new covering NC is a subdivision of C. Denote NG the symbolic image for a covering NC. The cells of the covering NG are denoted as $m(i, k)$, such that $\cup_k m(i, k) = M(i)$. There exists a multivalued mapping s from a graph NG to a graph G, such that a vertex (i, k) in NG is mapped to i in G, and a directed edge $(i, k) \to (j, l)$ is mapped to a directed edge $i \to j$. In this case, an admissible path in NG is mapped to an admissible path in G. In particular, periodic paths in NG are mapped into periodic paths in G.

Then for the sets

$$P = \{\cup M(i) : i \text{ is recurrent vertex on } G\},$$

$$NP = \{\cup m(i, k) : (i, k) \text{ is recurrent vertex on NG}\}$$

the inclusion $NP \subset P$ takes place. Applying this result at each step of the subdivision process, we obtain sequence of the nested neighborhoods

$$P_0 \supset P_1 \supset P_2 \supset \ldots \supset CR$$

of the chain-recurrent set.

2. Let d_k be the diameter of the covering C_k. If d_k tends to zero, then

$$\bigcap_k P_k = CR.$$

where *CR* is the chain-recurrent set. The inclusions established earlier hold for every symbolic image. Therefore, as k increases, the neighborhoods P_k converge to the trajectory *CR*. This establishes the theorem, confirming both properties of the sequence P_k.

Corollary. *Let Ω be a component of CR, then*

$$\{\cup M(i) : i \in H(\Omega)\} = P(\Omega) \tag{3.8}$$

is a closed neighborhood of the component Ω.

If $d_k \to 0$, *then*

$$\lim_{k\to\infty} P_k(\Omega) = \bigcap_k P_k(\Omega) = \Omega.$$

Example 3.1. *Localization of a chain-recurrent set of van der Pol differential equation.*

The van der Pol oscillator serves as a model of a triode tube oscillator in the case of a nonlinear tube characteristic [6]. The governing equation is

$$\ddot{x} - 1.5\dot{x}\left(1 - x^2\right) + x = 0.$$

Transitioning to a two-dimensional system, we obtain an autonomous system of differential equations:

$$\begin{aligned}\dot{x} &= y,\\ \dot{y} &= 1.5\dot{y}\left(1 - x^2\right) - x.\end{aligned}$$

To construct a discrete system, we create a shift operator Φ(x,y) per unit time along the trajectories of this system using the numerical Runge–Kutta method. We consider the resulting discrete dynamic system

$$(x_{n+1}, y_{n+1}) = \Phi(x_n, y_n).$$

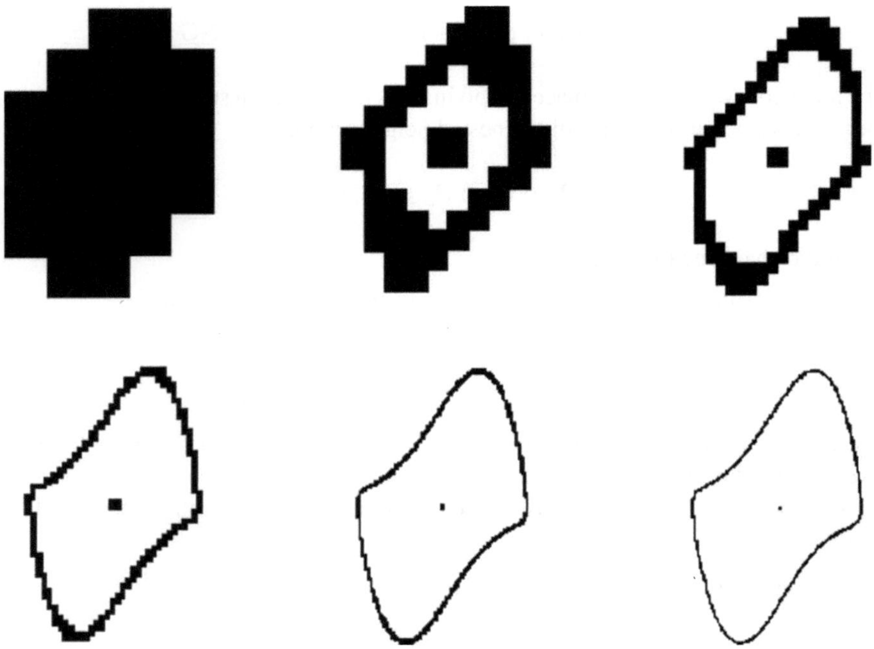

Fig. 3.1 Localization of the chain-recurrent set of van der Pol equation

We find the chain-recurrent set of the resulting system using the method described earlier. It is known that the van der Pol equation has an unstable equilibrium state at (0,0) and a stable periodic trajectory. There are no other chain-recurrent trajectories. Figure 3.1 illustrates the sequential construction of the sets $P_k, k = 1, 2, 3, 4, 5, 6$. It is evident that the set P_6 represents a sufficiently small neighborhood of a chain-recurrent set.

Example 3.2. Localization of a Chaotic Set.

A component of a chain-recurrent set Ω is called a chaotic invariant set if it contains a dense non-periodic trajectory. Japanese researchers, led by Ikeda (see [7, 8]), described the dynamics of a nonlinear optical system using a discrete dynamic system in the complex plane. The Ikeda mapping has the following form:

$$J : z \to R + C_2 z exp\left(i\left(C_1 - \frac{C_3}{1 + |z|^2}\right)\right),$$

where z is a complex variable, and R, C_1, C_2, C_3 are real constants. The Ikeda mapping models rather complex dynamics on the plane (see [9]). Ikeda demonstrated that the dynamical system described by this mapping as $R = 0.9, C_1 = 0.4, C_2 = 0.9, C_3 = 6$ can exhibit a chaotic attractor. To construct such an attractor, it is sufficient to generate a sufficiently long trajectory within the attraction set, which indicates where the attractor is located.

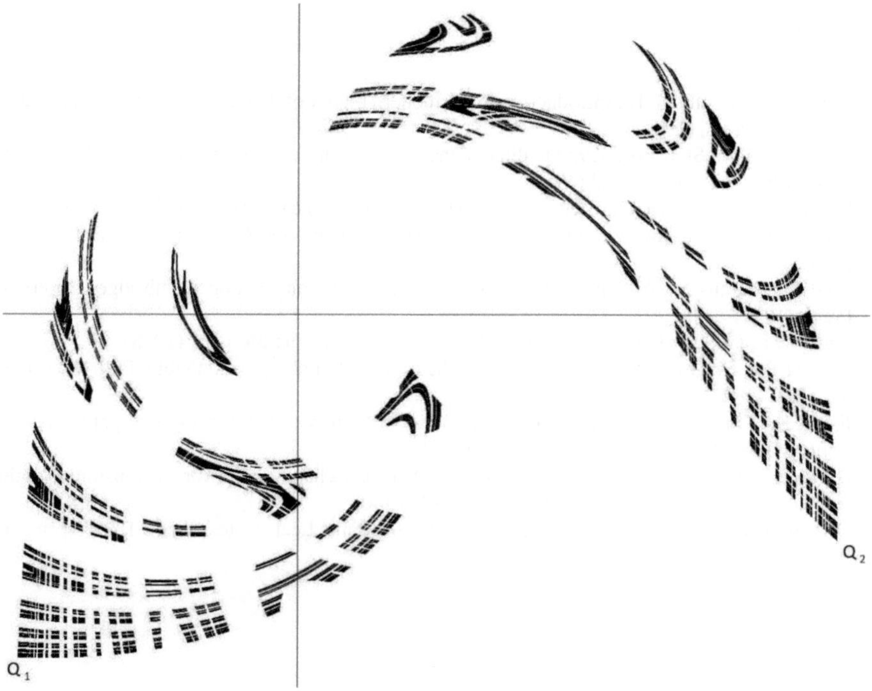

Fig. 3.2 Localization of the chaotic chain-recurrent set

The situation becomes more complex if the chaotic set Ω is not an attractor. In this case, we can apply the localization method described in Theorem 3.9, based on the principle of recurrence of pseudo-trajectories. Now, consider a modified Ikeda mapping of the form:

$$J : (x, y) \to (R + a(x \cdot cos\tau - y \cdot sin\tau, b(x \cdot sin\tau + y \cdot cos\tau)),$$

where $\tau = 0.4 - {}^{6}/_{(1+x^2+y^2)}$, $R = 1, a = -0.9, b = 1.2$. This mapping has a hyperbolic fixed point $H(-0.0950, 2.1937)$. There exists a 2-periodic hyperbolic trajectory $Q = \{(-1.5584, -1.9046), (3.0088, -1.2438)\}$, generating a homoclinic trajectory and a chaotic chain-recurrent set Ω. The set Ω forms a Cantor set, localized by the method described in Theorem 3.9.

The resulting neighborhood $P(\Omega)$ consists of 874,618 cells, as illustrated in Fig. 3.2. The set Ω comprises two parts: Ω_1 as the left part and Ω_2 as the right part. The Ikeda mapping is 2-periodic on Ω,

$$J : \Omega_1 \to \Omega_2; \Omega_2 \to \Omega_1,$$

with distinct dynamical behaviors manifesting within each segment.

References

1. Katok, A., Hasselblatt, B.: Introduction to the modern theory of dynamical systems. Cambridge University Press, (1995)
2. Nemytski, V.V., Stepanov, V.V.: Qualitative theory of dynamical systems. Princeton University Press, 1960. Russian original (1949)
3. Katok, A.B., Sinai, Ya. G., Stepin, A.M.: Theory of dynamical systems and general transformation groups with invariant measures. Itogi Nauki i Tekhn. Ser. Mat. Anal., **13**, 129–262 (in Russian) (1975)
4. Lind, D., Marcus, B.: An introduction to symbolic dynamics and coding. Cambridge University Press, (1995)
5. Shub, M.: Stabilite globale de systemes denamiques. Asterisque **56**, 1–21 (1978)
6. Van der Pol, B.: On relaxation-oscillations. The London, Edinburgh and Dublin Phil. Mag. J. of Sci., **2**(7) (1927), 978–992
7. Ikeda, K.: Multiple-valued stationary state and its instability of the transmitted light by a ring-cavity system. Opt. Comm. **30**, 257–261 (1979)
8. Ikeda, K., Daido, H., Akimoto, O.: Optical turbulence: chaotic behavior of transmitted light from a ring cavity. Phys. Rev. Lett. **45**, 709–712 (1980)
9. Osipenko G.: Dynamical systems, graphs, and algorithms. Lect. Notes Math., 1889, Springer, Berlin, (2007)

Chapter 4
Attractors and Filtrations

4.1 Definitions and Examples

In the study of dynamic systems, the invariant sets with properties of stability are of significant interest. Let $f : M \to M$ be a homeomorphism that generates a discrete dynamical system on the manifold M.

Definition 4.1 *A set $\Lambda \subset M$ is termed invariant if for any $x \in \Lambda$, both $f(x) \in \Lambda$ and $f^{-1}(x) \in \Lambda$.*

An invariant set Λ is classified as an attractor if there exists a neighborhood U such that

$$f\left(\overline{U}\right) \subset U, \Lambda = \bigcap_{n>0} f^n(U),$$

where the set

$$W^s(\Lambda) = \bigcup_{n<0} f^n(U)$$

is referred to as the basin or domain of attraction of Λ.

It is important to note that additional conditions may sometimes be incorporated into the definition of an attractor. An open set U with the property $f(\overline{U}) \subset U$ is termed an absorbing set, indicating that trajectories entering U remain within it. Thus, the attractor possesses an absorbing neighborhood, within which Λ is identified as the maximal invariant set. The set U is referred to as the fundamental neighborhood of the attractor U. We will demonstrate that by utilizing a symbolic image, it is possible to

G. Osipenko, *Symbolic Analysis of Dynamical Systems*, Mathematical Marvels: Texts and Monographs in the Spirit of CR Rao, https://doi.org/10.1007/978-981-95-0713-9_4

construct a sufficiently small absorbing neighborhood for any attractor and estimate the domain of its attraction.

Definition 4.2 *An invariant set* Λ^* *is referred to as a repeller for* f *if* Λ^* *serves as an attractor for* f^{-1}. *If* Λ *is an attractor with its fundamental neighborhood* U *and* $W^s(\Lambda)$ *as its basin, then the set.*

$$R = M \backslash W^s(\Lambda)$$

is identified as the repeller dual to Λ. It can be demonstrated that the set $M\backslash$U functions as a fundamental neighborhood of R for the mapping f^{-1}. In contemporary research, knowing the location of the attractor and its basin is of great importance. The basin is an open set, and every trajectory of a point within the basin eventually reaches a small fundamental neighborhood of the attractor in a finite number of iterations.

Example 4.1 (*Strange Attractors*). There exist many attractors with such non-trivial structures that they are classified as "strange attractors." This designation indicates that the attractor exhibits chaotic behavior, characterized by a dense orbit and a complex fractal structure, akin to the Cantor set. Henon [1] numerically explored the mapping given by

$$(x_1, y_1) = \left(1 + y - ax^2, bx\right).$$

It is known (see Fig. 4.1) that the dynamic system generated by the aforementioned mapping with parameters a = 1.4 and b = 0.3 within the domain [–1.5,1.5] × [–1.1] exhibits a strange attractor where the dynamics are chaotic. The Hausdorff dimension of the Henon attractor is approximately 1.26, as referenced in [2]. The Henon mapping is area contracting because the absolute value of its Jacobian is less than 1. Detailed numerical studies of the attractors of Henon mappings can be found in the book [3].

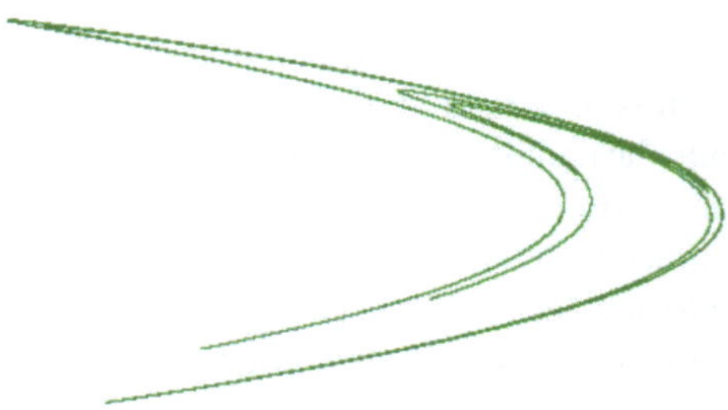

Fig. 4.1 The Henon attractor

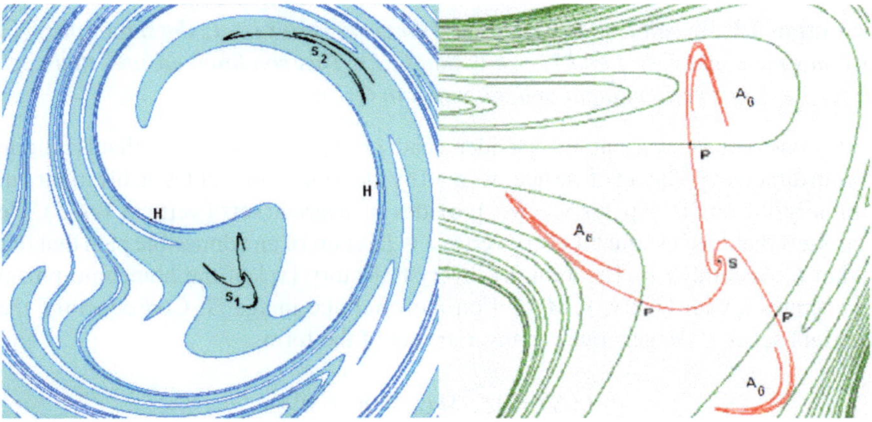

Fig. 4.2 The 2-periodic attractor "triple hook" and its basin

Example 4.2 The following example illustrates that the area of attraction can exhibit considerable complexity. Consider the Ikeda mapping defined by

$$f : (x, y) \to (R + C_2(xcos\tau - ysin\tau), C_2(xsin\tau + ycos\tau)),$$

where $\tau = C_1 - C_3/\left(1 + x^2 + y^2\right)$. Numerical simulations of the dynamics were conducted with the parameters $C_1 = 0.4$, $C_2 = 0.9$, $C_3 = 6.0$, and$R = 0.6$. The system possesses a 2-periodic attractor $A = S_1 + S_2$, where each component takes the form of a triple hook (see Fig. 4.2).

The property of two-periodicity implies that$f : S_1 \to S_2$ and$f : S_2 \to S_1$. The left part of Fig. 4.2 illustrates a portion of the domain of attraction, which is also 2-periodic. Each hook (as depicted in the right part of Fig. 4.2) resembles a Henon attractor, and together they constitute a periodic "strange" attractor. The boundary of the attraction domain corresponds to the unstable manifold of the hyperbolic 2-periodic trajectory H with coordinates (1.0094, −0.1100) and (−0.2110, −0.4211). The centers of each triple hook, S_1 and S_2, form a 2-periodic stable trajectory (refer to the right part of Fig. 4.2).

4.2 Attractor on Graph

Consider a symbolic image G corresponding to a covering with a maximal cell diameter d. Let $Ver(G)$ denote the set of vertices of the graph G. We can define a subgraph $G(L)$ with vertices $L \subset Ver(G)$ and edges $i \to j$ such that $i, j \in L$ and $i \to j \in G$.

Definition 4.3 *We call L an invariant set if for every vertex $i \in L$, the subgraph $G(L)$ includes the edges $j \to i$ and $i \to k$ within $G(L)$. This condition ensures that each vertex $i \in L$ has both an input and an output in $G(L)$.*

To construct an invariant set, consider a path $\omega = \{..., i_{-1}, i_0, i_1, ...\}$ that is infinite in both directions. The set of vertices formed by the path ω constitutes an invariant set. Similarly, the family of paths $S = \{\omega\}$ provides an invariant set of vertices $Ver(S)$. We can assert that L is invariant if every vertex $i \in L$ is part of an admissible path that lies within L. According to Theorem 2.1, every trajectory $\{x_k\}$ of the homeomorphism f generates a path $\{z_k : x_k \in M(z_k)\}$ on the symbolic image G. Consequently, the invariant set $\Lambda \subset M$ generates an invariant set of the form

$$L(\Lambda) = \{z : M(z) \cap \Lambda \neq \emptyset\}.$$

In particular, the set of all vertices $Ver(G)$ is invariant.

Let L be an invariant set of vertices on a symbolic image. We define the set of vertices

$$En(L) = \{i : i \notin L, \quad \text{there exists an edge } i \to j, \quad j \in L\}$$

as the entrance to L. Conversely, the set of vertices

$$Ex(L) = \{i : i \notin L, \quad \text{there exists an edge } j \to i, \quad j \in L\}$$

is termed the exit from L.

Definition 4.4

1. *We say that an invariant set of vertices L is an attractor if $Ex(L) = \emptyset$.*
2. *We say that an invariant set of vertices L is a repeller if $En(L) = \emptyset$.*

Let L be an attractor. The basin or domain of attraction is defined as the collection of vertices

$$D(L) = \{j : \textit{each path through } j \textit{ ends in } L\},$$

meaning that for every path $\{\cdots, j, \cdots, i_k, \cdots\}$, there exists a number k^* *such that the vertices* i_k with $i_k, k > k^*$ belong to L.

Proposition 4.1 [4]. Let L be an attractor and let $D(L)$ be its domain of attraction. Then:

1. The vertices from $D(L)\backslash L$ are nonrecurrent.
2. The set of vertices $L^* = Ver(G)\backslash D(L)$ is a repellor.

The repeller $L^* = Ver(G)\backslash D(L)$ *is called dual to the attractor L, and the pair L,* L^* is referred to as an attractor–repeller pair. The following statement describes the structure of an attractor on a symbolic image.

Proposition 4.2 [4]. *Each attractor is a union of classes of equivalent recurrent vertices and the paths connecting these classes.*

4.3 Attractors of a Dynamic System and Graph

As expected, there is a natural correlation between the attractors of a dynamical system and those of its symbolic image.

Theorem 4.1 [4]. *If L and D(L) are an attractor and its domain of attraction on a symbolic image, respectively, then there exists an attractor Λ of the homeomorphism f and its basin $W^s(\Lambda)$ such that a fundamental neighborhood of Λ takes the form*

$$U = Int\left\{\bigcup M(i), i \in L\right\},$$

where Int(.) denotes the interior. The set

$$W = \left\{\bigcup M(j), j \in D(L)\right\}$$

serves as an approximation of the basin $W^s(\Lambda)$ such that

$$\Lambda \subset U \subset W \subset W^s(\Lambda).$$

The following theorem indicates that both the attractor of a dynamical system and its domain of attraction can be constructed with high accuracy using a symbolic image, provided the diameter of its covering is sufficiently small, see Fig. 4.3.

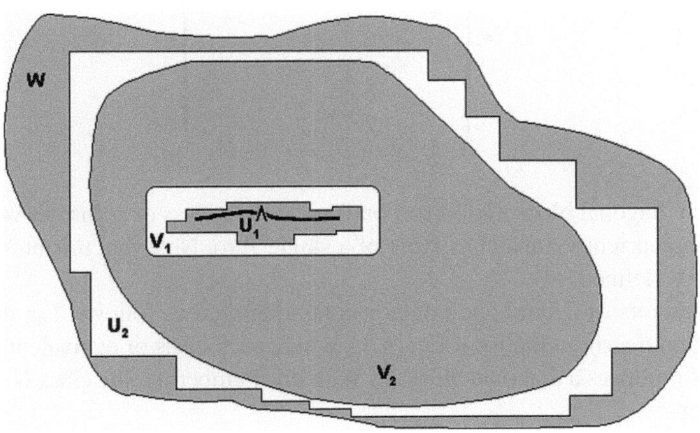

Fig. 4.3 Evaluation of the attractor and its area of attraction

Theorem 4.2 [4]. *Let $\Lambda \subset M$ be an attractor, V_1 be an arbitrarily small neighborhood of Λ, and V_2 be an arbitrarily large neighborhood such that*

$$\Lambda \subset V_1 \subset V_2 \subset \overline{V_2} \subset W^s(\Lambda).$$

Then, there exists a $d_0 > 0$ such that each symbolic image G with a maximal diameter of covering cells $d < d_0$ possesses an attractor L and a domain of attraction $D(L)$ satisfying

$$\Lambda \subset \left\{\bigcup M(i), i \in L\right\} \subset V_1 \subset V_2 \subset \left\{\bigcup M(j), j \in D(L)\right\} \subset W^s(\Lambda).$$

4.4 Transition Matrix and Attractors

Between the vertices of the symbolic image, there exists a quasi-order relation $i \longmapsto j$, indicating that there is a path of the form

$$i = i_0, i_1, i_2, ..., i_m = j.$$

According to Proposition 2.2, the vertices of the symbolic image can be renumbered such that equivalent recurrent vertices are assigned consecutive numbers. New indices for vertices i, j are chosen such that $i < j$ if i $\longmapsto$ j, with no corresponding path $j \longmapsto i$. The transition matrix then takes the form

$$\Pi = \begin{pmatrix} \Pi_1 & * & * & & \cdots * * \\ 0 & \ddots & * & & \cdots * * \\ 0 & 0 & \Pi_k & & \ddots \vdots \vdots \\ \vdots & \ddots & \ddots & \ddots & * \;\; \vdots \\ \vdots & \ddots & \ddots & \ddots & \ddots \;\; * \\ 0 & 0 & \cdots & \cdots & 0 \;\; \Pi_s \end{pmatrix}, \tag{4.1}$$

where each diagonal block Π_k corresponds either to a class of recurrent vertices H_k or to a nonrecurrent vertex, consisting of a single zero. Note that this numbering is not uniquely defined.

The attractors of a symbolic image can be identified as follows. Let us express the transition matrix in the form Eq. (4.1), where each class of equivalent recurrent vertices H produces a diagonal block Π. With this numbering, the class H generates an attractor

$$L = \{i \geq \min(j \in H)\}$$

because L has no exit. Similarly, the class H generates the repeller

$$R = \{i \leq \max(j \in H)\}$$

because R has no entrance. The intersection $R \cap L$ corresponds to class H. Two neighboring non-zero blocks Π_k and Π_{k+1} of the matrix Eq. (4.1) generate an attractor

$$L = \{i \geq min(j \in H_{k+1})\}$$

and the dual repeller

$$L^* = \{i \leq max(j \in H_k)\}$$

The basin of the attractor L is

$$D(L) = \{i > \max(j \in H_k)\}$$

since all vertices in $D(L)\backslash L$ are nonrecurrent. From Propositions 2.2, 4.1, and 4.2, it follows that for any attractor $L = \{i\}$, its domain of attraction $D(L)$ and the corresponding repeller $L^* = \{k\}$ can be renumbered such that k < j < i, where $j \in D(L)\backslash L$. The vertices from the set $D(L)\backslash L$ are nonrecurrent. Therefore, the transition matrix takes the form

$$\Pi = \begin{pmatrix} \Pi_1^* & * & * & \cdots & * & * \\ 0 & 0 & * & \cdots & * & * \\ 0 & 0 & \ddots & \ddots & \vdots & \vdots \\ \vdots & \ddots & \ddots & \ddots & * & \vdots \\ \vdots & \ddots & \ddots & \ddots & 0 & * \\ 0 & 0 & 0 & 0 & 0 & \Pi_2^* \end{pmatrix}, \tag{4.2}$$

where the blocks Π_1^* and Π_2^* correspond to the repeller L^* and the attractor L, respectively.

4.5 Filtration

A dynamical system typically possesses numerous attractors and repellers. The dynamics of an attractor–repeller pair dictate that a trajectory starting outside both the attractor and the repeller proceeds from the repeller and converges to the attractor. When a system contains multiple attractors, this results in complex dynamics. In such cases, the concept of filtration proves to be highly useful. Filtration serves as a discrete analogue to the Lyapunov function and exhibits stability with respect to perturbations.

Let f be a homeomorphism on a manifold M.

Definition 4.5 *A filtration for the homeomorphism f is defined as a finite sequence $F = \{F_0, F_1, \cdots, F_m\}$ of open sets such that.*

$$\emptyset = F_0 \subset F_1 \subset \cdots \subset F_m = M$$

and

$$f\left(\overline{F_k}\right) \subset F_k$$

for each $k = 0, 1, ..., m$.

The second condition is termed the absorbing property. The following proposition elucidates the structure of attractors generated by a filtration.

Proposition 4.3 [5]. *For a given filtration F, the maximal invariant subset in F_k*

$$A_k = \left\{\bigcap f^n(F_k) : n \in N\right\}$$

is an attractor, and the attractors A_k are nested within one another:

$$\emptyset = A_0 \subset A_1 \subset \cdots \subset A_m = M. \tag{4.3}$$

From this proposition, it follows that each F_k constitutes the fundamental neighborhood of the attractor A_k. Each attractor A_k has a corresponding dual repeller R_k. The intersection of an attractor and a repeller is known as a Morse set. The maximal invariant set I_k in $F_k \backslash F_{k-1}$ represents the intersection of the attractor A_k and the dual repeller R_{k-1} of A_{k-1}:

$$I_k = A_k \cap R_{k-1}.$$

The finite collection $\{I_k\}$ is referred to as the Morse decomposition.

Example 4.3 Consider a diffeomorphism that represents a shift of one unit in time along the trajectories of the following system of differential equations:

$$\dot{x} = x - x^3,$$
$$\dot{y} = -y$$

on the plane.

This system forms a discrete dynamical system characterized by three equilibrium points: $(-1, 0)$, $(0, 0)$, and$(1, 0)$. The equilibrium at (0,0) is a hyperbolic point, while the equilibria at $(\pm 1, 0)$ function as attractors. Let U_1 denote a fundamental neighborhood of $(-1, 0)$ and V_2 be a fundamental neighborhood of (1,0). We define $U_2 = U_1 \cup V_2$. Furthermore, there exists an open set U_3 that contains the neighborhoods U_1 and V_2, as well as the rectangle $[-1, 1] \times [-\varepsilon, \varepsilon]$, such that each trajectory intersects the boundary of U_3 transversely, entering its interior (see Fig. 4.4). This results in the filtration:

$$\emptyset \subset U_1 \subset U_2 \subset U_3 \subset R^2.$$

Notably, we can swap $(-1, 0)$ with $(1, 0)$ due to symmetry. The attractor A_1 consists of the single point $(-1, 0)$, while attractor A_2 comprises the two points $(-1, 0)$ and $(1, 0)$. The attractor A_3 is represented by the interval $[-1, 1] \times 0$. This leads to the inclusions:

$$\emptyset = A_0 \subset A_1 \subset A_2 \subset A_3 \subset A_4 = R^2.$$

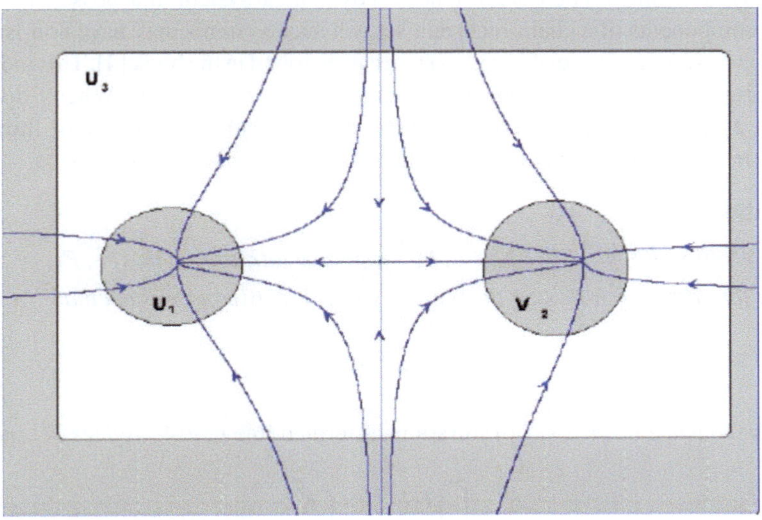

Fig. 4.4 The filtration $\emptyset \subset U_1 \subset U_2 = U_1 \cup V_2 \subset U_3 \subset R^2$

The next proposition demonstrates that a filtration is preserved under perturbations of the system.

Proposition 4.4 [5]. *If a sequence F is a filtration for f, then there exists a neighborhood V of f in the C^0-topology such that the sequence F is also a filtration for any map $g \in V$.*

Let $F = \{F_0, F_1, \cdots, F_m\}$ be a filtration. The maximal invariant subset in $F_k \backslash F_{k-1}$ is a Morse set defined as

$$I_k(F) = \left\{ \bigcap f^n(F_k \backslash F_{k-1}) : n \in Z \right\}.$$

We can define $I(F) = \{\cup I_k(F) : k = 1, \cdots, m\}$.

Proposition 4.5 [4]. *The chain-recurrent set is included in $I(F)$:*

$$CR \subset I(F) = \bigcup_k I_k(F).$$

A filtration F is termed fine if $I(F)$ coincides with the chain-recurrent set. In this scenario, the chain-recurrent set consists of a finite number of components:

$$CR = \bigcup_k \Omega_k, \Omega_k = I_k(F)$$

with the set $F_k \backslash \overline{F_{k-1}}$ acting as an isolating neighborhood for Ω_k. Generally, certain conditions must be met to ensure that the number of chain-recurrent components is finite. An illustrative example shows that even if a fine filtration exists, it is not uniquely determined. There are instances of dynamical systems that possess infinitely many components of a chain-recurrent set, yet such systems may have non-isolated components. An example of such a system can be found in the book [4]. This indicates that there are dynamical systems for which no fine filtration exists. When f does not permit a fine filtration, it is common practice to consider a sequence of filtrations $\{F_l : l = 1, 2, \ldots\}$ for which the sequence $I(F_l)$ approaches CR as $l \to \infty$.

Definition 4.6

(1) *Filtration* $F^* = \left\{F_0^*, F_1^*, \cdots, F_p^*\right\}$ *refines a filtration* $F = \{F_0, F_1, \cdots, F_m\}$, *if for each* $\alpha = 1, \cdots, p$ *there exists* $\beta(\alpha)$, $1 \le \beta(\alpha) \le m$ *such that*

$$F_\alpha^* \backslash F_{\alpha-1}^* \subset F_{\beta(\alpha)} \backslash F_{\beta(\alpha)-1}.$$

(2) A *sequence* $F^1, F^2, \cdots$ of filtrations is termed fine *if* F^{k+1} *refines* F^k *and*

$$\bigcap_k I(F^k) = CR.$$

The filtration sequence controls the growth of the chain-recurrent set. We will demonstrate that for any homeomorphism f, there exists a fine filtration sequence that can be constructed using a special sequence of symbolic images generated by the subdivision process.

4.6 Filtration on Symbolic Image

Definition 4.7 A finite sequence $\Phi = \{B_0, B_1, \cdots, B_m\}$ of vertex sets of a symbolic image G is called a filtration if.

$$\emptyset = B_0 \subset B_1 \subset \cdots, \subset B_m = Ver(G),$$

and B_k *has no exit from* B_k.

Proposition 4.6 [4]. *Each maximal invariant set* $L_k \subset B_k$ *is an attractor, yielding the inclusion*

$$\emptyset = L_0 \subset L_1 \subset \cdots, \subset L_m = Ver(G).$$

From Theorem 4.1, it follows that for each k, the set B_k generates a fundamental neighborhood

$$F_k = Int\Big\{\bigcup M(z), z \in B_k\Big\}$$

of an attractor A_k. Consequently, the collection of open sets

$$\emptyset = F_0 \subset F_1 \subset \cdots \subset F_m$$

forms a filtration on M. Thus, each filtration of a symbolic image generates a corresponding filtration of a dynamical system.

Let $\Phi = \{B_0, B_1, \cdots, B_m\}$ be a filtration on a symbolic image. The maximal invariant subset of the set $B_k \backslash B_{k-1}$ is denoted by $J_k(\Phi)$, and we define

$$J(\Phi) = \bigcup_k J_k(\Phi).$$

The set of recurrent vertices, denoted RV, lies within $J(\Phi)$:

$$RV \subset J(\Phi).$$

We will call a filtration Φ fine if the recurrent vertices satisfy $RV = J(\Phi)$. We denote the classes of equivalent recurrent vertices by H_p, where $p = 1, ..., s$. We

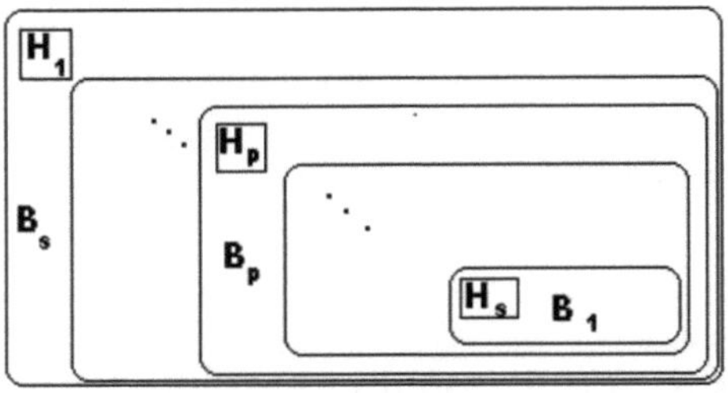

Fig. 4.5 Filtration on a symbolic image

write $H_p \longmapsto H_q$ if and only if there is a valid path from H_p to H_q. Let the vertices of the symbolic image be numbered according to Proposition 2.2. In this case, the transition matrix takes the form:

$$\Pi = \begin{pmatrix} \Pi_1 & * & * & \cdots ** \\ 0 & \ddots & * & \cdots ** \\ 0 & 0 & \Pi_p & \ddots \vdots \vdots \\ \vdots & \ddots & \ddots & \ddots \; * \; \vdots \\ \vdots & \ddots & \ddots & \ddots \; \ddots \; * \\ 0 & 0 & \cdots & \cdots \; 0 \; \Pi_s \end{pmatrix}$$

where the elements under the diagonal blocks are zeros, each diagonal block Π_p corresponds to either a class of equivalent recurrent vertices H_p or a nonrecurrent vertex. In the last case Π_p coincides with zero. Let us introduce the numbers $n(H_p) = min\{i : i \in H_p\}$. Each class H_p generates an attractor of the form

$$E_p = \{i : i \geq n(H_p)\}, p = 1, ..., s.$$

Let us put $B_k = E_p$, where $p = s + 1 - k$, $k = 0, 1, ..., s$. We have $B_0 = \emptyset$ and $B_s = Ver(G)$. Figure 4.5 illustrates this design.

Proposition 4.7 [4]. *The constructed sequence* $\Phi = \{B_0, B_1, \cdots, B_s = Ver(G)\}$ serves as *a fine filtration of the symbolic image.*

Now we are ready to apply the process of successive subdivision of coverings to obtain a fine sequence of filtrations of a dynamic system.

Algorithm for construction of a fine sequence of filtrations.

(1) Let C be an arbitrary finite covering of M by closed cells. Based on this covering, the symbolic image G is constructed.
(2) Classes H_p of equivalent recurrent vertices are identified. We introduce a canonical numbering for the vertices. Let $d = max\{diamM(i) : i \text{ is recurrent}\}$.
(3) A filtration $\Phi = \{B_0, B_1, \cdots, B_s = Ver(G)\}$ of the symbolic image is obtained by setting $B_k = \{i : i \geq min\{i : i \in H_p\}\}, p = s + 1 - k\}$.

(4) The filtration $F = \{F_0, F_1, \ldots, F_s\}$ for the dynamic system is defined by setting $F_k = \{\cup M(i) : i \in B_k\}$.
(5) Cells corresponding to the recurrent vertices are subdivided and a new covering is determined.
(6) The symbolic image for the new covering is constructed.
(7) Return to the second step.

The described algorithm generates a sequence of symbolic images G_m, filtrations Φ_m on each G_m, a sequence of filtrations F_m on M and the sequence of numbers d_m. The following theorem justifies the proposed algorithm.

Theorem 4.3 [4]. *If $d_m \to 0$, then the sequence of filtration F_m is fine.*

Corollary *For each homeomorphism f there exists a fine sequence of filtrations.*

Let G be a symbolic image with respect to the covering C, and let d be the diameter of the covering C. The multivalued mapping $h : M \to V$ associates to a point x all the vertices i such that $x \in M(i)$. Let Ω be a component of a chain-recurrent set. The image $h(\Omega)$ lies within some class of equivalent recurrent vertices $H(\Omega)$, and the component Ω lies in the union $\{\cup M(i) : i \in H(\Omega)\}$. From Proposition 4.7, it follows that the class $H(\Omega)$ is the intersection of some attractor L^+ and repeller L^- on the symbolic image. According to Theorem 4.1, an attractor on a symbolic image generates a fundamental neighborhood

$$U^+ = Int\{\cup M(i), i \in L^+\}$$

of some attractor A^+, which contains the neighborhood $\{\cup M(i), i \in H(\Omega)\}$ of the component Ω. Similarly, the repeller L^- on the symbolic image generates a fundamental neighborhood

$$U^- = Int\{\cup M(i), i \in L^-\}$$

of some repeller A^-, which also contains the neighborhood $\{\cup M(i), i \in H(\Omega)\}$ of the component Ω. The intersection $A^+ \cap A^-$ forms a Morse set that contains the component Ω. According to Theorem 4.2, attractors and repellers can be constructed with arbitrary precision. Let us assume the component Ω is isolated, meaning there exists a neighborhood V such that Ω is the largest invariant set within V. In this case, the set $\{\cup M(i) : i \in H(\Omega)\}$ lies within V for a sufficiently small covering diameter d. Thus, we arrive at the following statement:

Proposition 4.8 An isolated component of the chain-recurrent set is the intersection of an attractor and a repeller; that is, an isolated component is a Morse set.

References

1. Hénon, M.: A two-dimensional mapping with strange attractor. Comm. Math. Phys. **50**, 69–77 (1976)
2. H. G. Schuster and W. Just. Deterministic Chaos: An Introduction. Wiley-VCH, Weinheim (2005).
3. Nusse, H., Yorke, J.: Dynamics: numerical exploration. Springer New-York Berlin (1997)
4. Osipenko, G.: Dynamical systems, graphs, and algorithms. Lect. Notes Math., 1889, Springer, Berlin, (2007)
5. Bronshtein, I.U.: Non-autonomous dynamical system. Kishiniv, Shtinitsa, (1984) (in Russian)

Chapter 5
Invariant Measures

5.1 Flows on Symbolic Image

Let $f : M \to M$ be a homeomorphism of a compact manifold M. In the following discussion, we will utilize the concepts of measurable sets, measures, Lebesgue integrals, and related topics. Necessary information on these subjects can be found in the Appendix chapter.

A measure μ on M is said to be invariant under the homeomorphism f if, for any measurable set $A \subset M$, it holds that

$$\mu\big(f^{-1}(A)\big) = \mu(A) = \mu(f(A)).$$

The Krylov–Bogolyubov theorem [1, 2] guarantees the existence of an invariant measure μ that is normalized on M, such that $\mu(M) = 1$. We denote the set of all f-invariant normalized measures by $\mathcal{M}(f)$. In this chapter, we demonstrate that any f-invariant measure μ and the set $\mathcal{M}(f)$ can be constructed using a sequence of symbolic images as the diameter of the covering converges to zero.

If μ_1, μ_2 are invariant measures, and $\alpha, \beta \geq 0$ with $\alpha + \beta = 1$, then the measure

$$\mu = \alpha\mu_1 + \beta\mu_2$$

is also an invariant measure [1]. Consequently, the set of all f-invariant measures $\mathcal{M}(f)$ is convex and, moreover, this set is compact in the weak topology, as noted in [3]. The convergence $\mu_n \to \mu$ in this topology means that

$$\int\limits_M \varphi d\mu_n \to \int\limits_M \varphi d\mu$$

for any continuous function $\varphi : M \to R$.

G. Osipenko, *Symbolic Analysis of Dynamical Systems*, Mathematical Marvels: Texts and Monographs in the Spirit of CR Rao, https://doi.org/10.1007/978-981-95-0713-9_5

A measure μ is ergodic if for any invariant measurable set A, it holds that either $\mu(A) = 0$ or $\mu(M \backslash A) = 0$. A point μ in the convex set $\mathcal{M}(f)$ is termed an extreme point if it cannot be expressed in the form:

$$\mu = \alpha\mu_1 + (1-\alpha)\mu_2,$$

where $\mu_1,\ \mu_2 \in \mathcal{M}(f)$ and $0 < \alpha < 1$. It is known [4] that the extreme points of the convex set $\mathcal{M}(f)$ are ergodic measures.

Example 5.1 Consider a point x that is periodic with the smallest period p. The trajectory $T = \{x, f(x), \cdots, f^{p-1}(x)\}$ consists of p distinct points, and we have $f^p(x) = x$. Let the measure μ be uniformly distributed along the trajectory T, meaning that $\mu(x) = \mu(f(x)) = \cdots = \mu\big(f^{p-1}(x)\big) = 1/p$. The measure of the set A is then defined by

$$\mu(A) = \mu(A \cap T) = k/p,$$

where k is the number of points in the trajectory T that lie within A. It is straightforward to verify that this measure is invariant and ergodic, and it is discrete in the sense that the measure is concentrated at a finite number of points.

Ulam [5] proposed a method for constructing a sequence of measures through the approximation of the Perron–Frobenius operator; such a sequence converges to an invariant measure in the weak topology. Articles [6, 7] present a method for constructing the invariant Sinai–Bowen–Ruelle measure (SBR measure) with favorable properties. The SBR measure is absolutely continuous with respect to the Lebesgue measure, which is why it is often referred to as "natural" or "physical." While not every mapping f possesses an invariant SBR measure, every mapping f does have a set of invariant measures $\mathcal{M}(f)$. Our objective is to construct the entire set of invariant measures $\mathcal{M}(f)$.

The proposed methodology enables the construction of an individual invariant measure, potentially endowed with additional properties. To address this challenge, we will utilize the symbolic image of a dynamic system. Let $C = \{M(1), M(2), \cdots, M(n)\}$ be a finite covering of M and let G denote the symbolic image. Without loss of generality, we can assume that the cells $M(i)$ are polyhedra that intersect at boundary disks. The general scheme is as follows: we define a probability distribution m_{ij} on the edges that adheres to certain invariance properties. This distribution is referred to as a flow on the symbolic image, and it serves as an approximation of an invariant measure.

To understand the proposed construction, let us assume that the mapping f has an invariant measure μ on the manifold M. We will construct a measurable partition $C^* = \{M^*(i)\}$, where boundary disks are attributed to one of the neighboring cells. We can then assign a measure to each arc $i \to j$ of the symbolic image via the formula:

$$m_{ij} = \mu\big(M^*(i) \cap f^{-1}\big(M^*(j)\big)\big) = \mu\big(f\big(M^*(i)\big) \cap M^*(j)\big), \tag{5.1}$$

where the second equality arises from the invariance of μ. Additionally, we observe that

$$\sum_k m_{ki} = \sum_k \mu\big(f\left(M^*(k)\right) \cap M^*(i)\big) = \mu\big(M^*(i)\big)$$
$$= \sum_j \mu\big(M^*(i) \cap f^{-1}\left(M^*(j)\right)\big) = \sum_j m_{ij}.$$

The sum $\sum_k m_{ki}$ can be referred to as the incoming flow to vertex i, while the sum $\sum_j m_{ij}$ denotes the outgoing flow from i. The equality

$$\sum_k m_{ki} = \sum_j m_{ij} \tag{5.2}$$

can be interpreted as a Kirchhoff-type law: the incoming flow is equal to the outgoing flow for each vertex. Moreover, we also have the equality

$$\sum_{ij} m_{ij} = \mu(M) = 1, \tag{5.3}$$

which indicates that the distribution m_{ij} *is normalized or probabilistic. Thus, the invariant measure* μ *generates a distribution* m_{ij} that satisfies conditions (5.2) and (5.3). This observation leads us to the following definition.

Definition 5.1 *Let G be a directed graph. A distribution* $\{m_{ij}\}$ *on the edges* $\{i \to j\}$ *is called a flow if it satisfies the following conditions:*

$$m_{ij} \geq 0;$$
$$\sum_{ij} m_{ij} = 1;$$
$$\sum_k m_{ki} = \sum_j m_{ij} \text{ for any vertex i}$$

The last property is referred to as flow invariance. The second normalization property can be rewritten as $m(G) = 1$, where the measure of a graph is defined as the sum of the measures of its arcs. In graph theory, the described distribution is sometimes called a closed or invariant flow.

For the flow $\{m_{ij}\}$ on the graph, we can define the measure of vertex i as follows:

$$m_i = \sum_k m_{ki} = \sum_j m_{ij}.$$

In this case, we obtain the equality $\sum_i m_i = m(G) = 1$. Thus, each invariant measure generates a flow on the symbolic image.

Now, let's consider the reverse construction. Let $m = \{m_{ij}\}$ be a flow on the symbolic image G. We can define a measure μ as follows:

The measure of a measurable set A can be defined by the formula:

$$\mu(A) = \sum_i \frac{m_i \nu(A \cap M(i))}{\nu(M(i))}, \tag{5.4}$$

where ν is the Lebesgue measure. It is assumed that the Lebesgue measure of each cell $\nu(M(i)) \neq 0$. By this definition, the measure of the cell $M(i)$ coincides with the measure of the vertex i:

$$\mu(M(i)) = m_i.$$

Since ν represents the Lebesgue measure, the constructed measure of boundary disks is zero, meaning that the measure of the cell $\mu(M(i)) = m_i$ does not depend on whether the boundary disks belong to that cell. Generally, the constructed measure μ is not invariant under f. However, as will be shown later, this measure approximates an invariant measure in the sense that μ converges in the weak topology to an invariant measure as the diameters of the coverings tend toward zero. Section 5.6 presents another method for constructing measures on manifolds. Within each cell M(i), we fix a point x(i) that does not lie within any other cells. We then consider the flow $\{\mathrm{m_{ij}}\}$ on G and define the vertex measure i as

$$m_i = \sum_k m_{ki} = \sum_j m_{ij}.$$

Now, we will construct a measure μ^* concentrated at the points $x(i)$:

$$\mu^* = \sum_i m_i \delta(x(i)),$$

where $\delta(x)$ is the standard delta measure concentrated at the point x. For any measurable set A, we define

$$\mu^*(A) = \sum_{x(i) \in A} m_i. \tag{5.5}$$

In Sect. 5.6, we show that the measure μ^* converges in the weak topology to an invariant measure if the diameter of the covering tends to zero.

5.2 Flow Properties

Let G be an abstract directed graph with n vertices. Consider the space of all flows $\mathcal{M}(G) = \{m\}$ on G. Let $m^1 = \left\{m^1_{ij}\right\}$ and $m^2 = \left\{m^2_{ij}\right\}$ be two flows, and let $\alpha, \beta \geq 0$ such that $\alpha + \beta = 1$. It is straightforward to verify that the distribution $m = \alpha m^1 + \beta m^2$ is also a flow. In this context, we say that the flow m is the convex combination of the flows m^1 and m^2 with weights α and β. Thus, the space of all flows $\mathcal{M}(G)$ is convex. Next, we will examine the structure of the set $\mathcal{M}(G)$. A periodic path

$$\omega = (i_0 \to i_1 \to i_2 \to \cdots \to i_k = i_0)$$

is termed simple or a cycle if the vertices $\{i_t, t = 1, 2, \cdots, k\}$ are distinct. The simple path ω generates a flow $m(\omega)$ located on the cycle ω such that $m_{ij} = 1/k$ for all edges of the periodic path ω, and $m_{ij} = 0$ for all other edges. The constructed flow $m(\omega)$ is termed a simple flow. It is evident that $m(\omega)$ is invariant, unique, and indecomposable into the sum of other flows. In other words, $m(\omega)$ is an extreme point of the set of flows $\mathcal{M}(G)$. Since the number of vertices is finite, then the number of cycles and the number of simple flows is also finite.

The Choquet theorem (see [1]) states that any point in a compact metrizable convex set can be expressed as an integral sum of its extreme points. The following theorem serves as an analogue of this theorem for the flows $\mathcal{M}(G)$ on a graph:

Theorem 5.1 *Any flow $m \in \mathcal{M}(G)$ can be decomposed into a sum of the simple flows:*

$$m = \sum_k \alpha_k m(\omega_k),$$

where $\alpha_k \geq 0$, $\sum_k \alpha_k = 1$, and $\{\omega_k\}$ is the complete set of cycles on G. The set $\mathcal{M}(G)$ is a convex polyhedron, with the simple flows serving as its vertices (extreme points).

Proof Let $m = \left\{m_{ij}\right\}$ be a flow on G. *Consider the set of edges (arcs) $D = \{i \to j\}$ such that $m_{ij} > 0$.* Kirchhoff's law (5.2) holds on this set. It follows that there exists a path ω^* of infinite length that passes through each edge in D. Specifically, if there exists an arc $k \to i$ with $m_{ki} > 0$, then, according to (5.2), there must be an arc $i \to j$ with $m_{ij} > 0$. Therefore, we can continue the path within D.

Since D consists of a finite number of edges, ω^* must eventually contain a periodic path ω, which can be considered simple. Let p be the minimal period of the cycle ω, and define:

$$m_{min} = min\left\{m_{ij} : i \to j \in \omega\right\} > 0,$$

which denotes the minimum measure of edges in ω. Let $\alpha > 0$ be defined such that

$$\alpha = pm_{min}\, or\, \alpha/p = m_{min}.$$

Now, we construct a new distribution m^* on the edges of G. For each edge $i \to j$ in ω, we define a new measure:

$$m^*_{ij} = m_{ij} - \frac{\alpha}{p} \geq 0.$$

If an edge $i \to j$ is not included in ω, then $m^*_{ij} = m_{ij}$. Clearly, the sum of measures of all edges is

$$\sum_{ij} m^*_{ij} = 1 - \alpha.$$

Next, we will verify that Kirchhoff's law (5.2) holds for the distribution m^*. If a vertex i does not lie on the cycle ω, then $m^*_{ij} = m_{ij}$ and $m^*_{ki} = m_{ki}$, ensuring that the equality (5.2) holds. Now, consider the case where i lies on the simple cycle ω. In this cycle, there exists an edge $k^* \to i$ that enters i and an edge $i \to j^*$ that exits i; all other edges from the cycle ω have no connection to the vertex i, since ω is defined as a simple cycle, see Fig. 5.1.

In this case, the left- and right-hand sides of the equality

$$\sum_{k} m_{ki} = \sum_{j} m_{ij} \tag{5.6}$$

contain, respectively, the terms m_{k^*i} and m_{ij^*} generated by the edges $k^* \to i$ and $i \to j^*$. All other terms have no connection with the cycle ω. Thus, we can express this as

$$m_{k^*i} + \sum_{k \neq k^*} m_{ki} = m_{ij^*} + \sum_{j \neq j^*} m_{ij}. \tag{5.7}$$

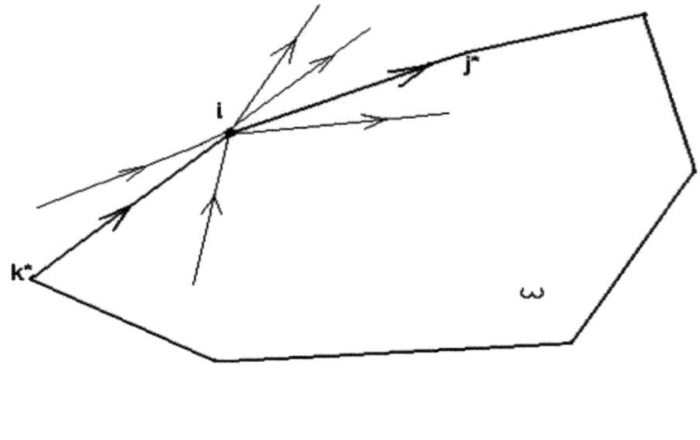

Fig. 5.1 Verification of Kirchhoff's law

By subtracting $\frac{\alpha}{p} = m_{min}$ from both sides of (5.7), we obtain

$$m_{k^*i} - \frac{\alpha}{p} + \sum_{k \neq k^*} m_{ki} = m_{ij^*} - \frac{\alpha}{p} + \sum_{j \neq j^*} m_{ij}$$

which can be rewritten as

$$\sum_k m^*_{ki} = \sum_j m^*_{ij}.$$

This demonstrates that the new distribution m^* satisfies Kirchhoff's law. Moreover, the set of arcs $D^* = \{i \to j, m^*_{ij} > 0\}$ does not include any edges from the cycle ω, since

$$m^*_{min} = m_{min} - \frac{\alpha}{p} = 0$$

on ω. The cycle ω generates a simple flow m_ω that is equal to $\frac{1}{p}$ on each edge of ω and is zero on all other edges. According to our construction,

$$m_{ij} = m^*_{ij} + \frac{\alpha}{p}$$

for edges from ω and

$$m_{ij} = m^*_{ij}$$

for other edges. This leads to the equality:

$$m = m^* + \alpha m_\omega.$$

By iteratively eliminating simple cycles from the set D, we achieve a zero distribution in a finite number of steps. Thus, the original flow can be represented in the form:

$$m = \sum_\omega \alpha_\omega m_\omega,$$

where $\alpha_\omega \geq 0$, m_ω is a simple flow, and the sum is taken over all simple cycles. From the equalities $m(G) = m_\omega(G) = 1$, it follows that

$$\sum_\omega \alpha_\omega = 1.$$

This completes the proof.

Comment. From the proof of Theorem 5.1, it follows that the measure m_{ij} can be positive only on the edge $i \to j$ when a periodic path passes through it; that is, the vertices i and j must be recurrent.

The set of recurrent vertices is decomposed into equivalence classes known as strong components. According to Theorem 3.9, the strong components of a symbolic image generate an isolated neighborhood of chain-recurrent sets of the dynamic system. It is known [1] that an invariant measure is equal to zero outside of a chain-recurrent set. Consequently, to construct invariant measures, it suffices to study an isolated invariant component of chain-recurrent sets. Thus, without loss of generality, we can assume that the graph G consists of a single strong component.

From Theorem 5.1, it follows that the family of flows forms a convex polyhedron, which is the convex hull of the simple flows. This means that any flow can be constructed using the following method. Let $\{\omega_z\}$ be the set of all simple paths (cycles) and $\{m(\omega_z)\}$ be the corresponding set of simple flows. Each simple flow $m(\omega_z) = \left\{m_{ij}^z\right\}$ is uniquely determined, and the set of all simple flows $\{m(\omega_z)\}$ is finite. According to Theorem 5.1, each flow $m = \left\{m_{ij}\right\}$ is characterized by a collection of values $a_z > 0$ such that

$$\sum\nolimits_z a_z = 1$$

and

$$m_{ij} = \sum\nolimits_z a_z m_{ij}^z.$$

The coefficients $\{a_z\}$ are referred to as the weights of the simple cycles $\{\omega_z\}$. Consequently, the flow m can be represented by a point in a standard simplex:

$$\Delta = \left\{a \in R^N : a_z \geq 0, \sum_z a_z = 1\right\},$$

where N is the number of simple cycles in G.

Using the method of construction based on all simple flows can require significant computation time, especially since the number of all simple cycles is generally vast. For instance, a complete graph with n vertices (where each vertex is connected to every other vertex) has n^2 edges and $N = 2^n$ simple cycles. To mitigate this, we can utilize a partial collection of cycles by assigning the weights of unused cycles to zero. Now, let us consider the general properties of flows on a graph.

Proposition 5.1 *Let Q and G be directed graphs and let $s : Q \to G$ be a mapping of directed graphs. Suppose that there exists a flow m on Q. Then the flow $m^* = s^* m$ is generated on G, where the measure of the edge $i \to j$ on G is evaluated as*

$$m_{ij}^* = \sum_{s(p \to q) = i \to j} m_{pq},$$

where the sum is taken over all edges $p \to q$ that are transformed into $i \to j$. If an edge $i \to j$ has no preimage, then $m^*_{ij} = 0$.

Proof To establish this, we need to verify two properties of a flow for m^*:

Normalization Property: The total measure across all edges in the graph G coincides with the total measure across all edges in Q:

$$\sum_{ij} m^*_{ij} = \sum_{pq} m_{pq} = 1.$$

Invariance Property: Fix a vertex $i \in G$:

If $s^{-1}(i) = \emptyset$, then all incoming and outgoing edges associated with i also lack preimages. This means that their measures equal zero, and thus the invariance condition holds. If $s^{-1}(i) \neq \emptyset$, consider all vertices p in the preimage $s^{-1}(i)$. The invariance condition is satisfied for each vertex p, leading to

$$\sum_{r} m_{rp} = \sum_{t} m_{pt}.$$

By summing these equalities over $p \in s^{-1}(i)$ and recognizing that $m^*_{rj} = 0$ for edges $r \to j$ with $s^{-1}(i \to j) = \emptyset$, we derive the necessary equality:

$$\sum_{k} m^*_{ki} = \sum_{j} m^*_{ij}$$

for $i \in G$. The proof is complete.

Proposition 5.2 *Let Q and G be directed graphs, let $s : Q \to G$ be a mapping of directed graphs, and suppose that there exists a flow m on Q, which can be expressed as a linear combination of flows:*

$$m = \alpha m_1 + \beta m_2, \quad \alpha \geq 0, \quad \beta \geq 0, \quad \alpha + \beta = 1.$$

Then,

$$s^*(m) = \alpha s^*(m_1) + \beta s^*(m_2),$$

indicating that s^* is a linear mapping.

Proof Let $m_1 = \left\{m^1_{pq}\right\}$ and $m_2 = \left\{m^2_{pq}\right\}$. Then, we have

$$m = \left\{m_{pq} = \alpha m^1_{pq} + \beta m^2_{pq}\right\}.$$

By definition,

$$s^*(m)_{ij} = m^*_{ij} = \sum_{s(p\to q)=i\to j} \left(\alpha m^1_{pq} + \beta m^2_{pq}\right) =$$
$$\alpha \sum_{s(p\to q)=i\to j} m^1_{pq} + \beta \sum_{s(p\to q)=i\to j} m^2_{pq}.$$

This can be expressed as follows:

$$m^*_{ij} = \alpha \sum_{s(p\to q)=i\to j} m^1_{pq} + \beta \sum_{s(p\to q)=i\to j} m^2_{pq} = \alpha m^{1*}_{ij} + \beta m^{2*}_{ij}$$
$$= \alpha\left(s^* m_1\right)_{ij} + \beta\left(s^* m_2\right)_{ij}.$$

Thus, the proof is complete.

Let us now introduce the distance metric in the space of flows. For flows $m_k = \left\{m^k_{ij}\right\}$ where $k = 1, 2$, we define the distance as follows:

$$\rho(m_1, m_2) = \sum_{ij} \left|m^1_{ij} - m^2_{ij}\right|.$$

In this topology, $\mathcal{M}(G)$ and $\mathcal{M}(Q)$ are convex compact sets.

Proposition 5.3 Let $\mathcal{M}(Q)$ be the set of flows on Q, and let $s : Q \to G$ be a mapping of directed graphs. Then the mapping s^* is continuous with respect to the metric ρ and $s^*(\mathcal{M}(Q))$ is a convex compact set in $\mathcal{M}(G)$.

The proof of this statement is straightforward, and we omit it for brevity.

Proposition 5.1 provides justification for the following methods of constructing flows on the graph G.

Proposition 5.4 *Let there be a periodic path ω of period N on a graph. Then there exists a flow m^* on G such that*

$$m^*_{ij} = \frac{k_{ij}}{N},$$

where k_{ij} is the number of times the path ω through the edge $i \to j$.

Proof. Let $\omega = \{i_1 \to i_2 \to \cdots \to i_N \to i_1\}$. Let us construct a graph Q consisting of one simple periodic cycle of length N; that is, $Q = \{1 \to 2 \to \cdots \to N \to 1\}$. In this case, there is a unique flow m on Q such that $m_{pq} = 1/N$. Let the mapping stack the cycle on the periodic path, that is, $s(k) = i_k$. According to Proposition 5.1, a flow is generated such that the measure of edge $i \to j$ is the sum of measures of edges in $s^{-1}(i \to j)$, in other words

$$m^*_{ij} = \frac{k_{ij}}{N},$$

where k_{ij} is the number of times the path ω through the edge $i \to j$. The proof is complete.

Let G consist of one strong component; then any two vertices or edges are connected by a permissible path. Therefore, there exists a periodic path passing through all the vertices. The path can be called dense on vertices. According to proposition 5.4, there is flow $m = \{m_{ij}\}$ on G with a positive measure $m_i = \sum_j m_{ij} > 0$ for each vertex. Since the graph G consists of one strong component, the proposition is true. ω^* passing through each edge; this path can be called dense on edges. Thus, on any graph, there exists a flow that is positive on each recurrent vertex or edge.

Proposition 5.5 *Let there be a family of periodic paths $\omega_1, \cdots, \omega_r$ on the graph G with periods $p_1, \cdots, p_r$. Define $N = p_1 + \cdots + p_r$. Then, there exists a flow m on G such that*

$$m_{ij} = \frac{k_{ij}}{N},$$

where k_{ij} is the number of passages of the paths $\omega_1, \cdots, \omega_r$ through the edge $i \to j$.

The proof of this proposition closely follows that of Proposition 5.4. Let us assume the hypotheses of the proposition. Note that Proposition 5.1 does not require the graphs G and Q to be connected. We can construct a graph Q that consists of a disconnected union of r simple periodic cycles $\Omega_1, \cdots, \Omega_r$ with periods $p_1, \cdots, p_r$. The graph Q has N vertices and N edges, where $N = p_1 + \cdots + p_r$.

It is straightforward to verify that there exists a flow m^* on Q with edge measure $m^*_{pq} = 1/N$. The mapping $s : Q \to G$ stacks the cycles $\Omega_1, \cdots, \Omega_r$ on the periodic paths $\omega_1, \cdots, \omega_r$, respectively. According to Proposition 5.4, the measure of the edge $i \to j$ is equal to the sum of the measures of its preimages, which means $m_{ij} = k_{ij}/N$ where k_{ij} is the number of passages of the paths $\omega_1, \cdots, \omega_r$ through the arc i→j. Thus, the proof is complete.

Proposition 5.5 admits the following generalization.

Proposition 5.6 *Let there be a family of periodic paths $\omega_1, \cdots, \omega_r$ on the graph G with periods $p_1, \cdots, p_r$ and associated numbers $\alpha_1, \cdots .\alpha_r$ such that $\alpha_t \geq 0$ $\sum_t \alpha_t = 1$. Then, there exists a flow m on G such that*

$$m_{ij} = \sum_{t=1}^{r} \alpha_t \frac{k^t_{ij}}{p_t},$$

where k^t_{ij} is the number of passages of the path ω_t through the edge $i \to j$.

Proposition 5.5 can be derived from Proposition 5.6 by setting $\alpha_t = p_t/N$, $N = \sum_t p_t$. From Theorem 5.1, it follows that any flow from $\mathcal{M}(G)$ can be expressed as described in Proposition 5.6.

Consider a directed graph G and the set of all flows $\mathcal{M}(G)$ on it. Let $m^1 = \left\{m^1_{ij}\right\}$ and $m^2 = \left\{m^2_{ij}\right\}$ lie in $\mathcal{M}(G)$. The distance $\rho\left(m^1, m^2\right) = max\left\{\left|m^1_{ij} - m^2_{ij}\right|\right\}$ is defined on the flow space $\mathcal{M}(G)$. The set of these flows $\mathcal{M}(G)$ is a convex compact space with respect to a metric ρ. Let H be the equivalence class of recurrent vertices. Then, the set $\mathcal{M}(H)$ of flows concentrated on H is also a convex compact set. Each periodic path ω generates a flow $m(\omega)$ according to Proposition 5.4. The flow is called a periodic flow.

Theorem 5.2 *Let H be an equivalence class of recurrent vertices, and let $\mathcal{M}(H)$ be the space of flows on H. Then the periodic flows are dense in $\mathcal{M}(H)$ with respect to the metric ρ.*

Proof. Consider a flow $m \in \mathcal{M}(H)$. According to Theorem 5.1, every flow can be decomposed into a sum of periodic flows. Suppose that m is decomposed into a sum of two periodic flows

$$m = \alpha m(1) + \beta m(2),$$

where $\alpha > 0, \beta > 0, \alpha + \beta = 1$, $m(1)$ and $m(2)$ are generated by the periodic paths ω_1 and ω_2. We claim that m can be approximated by periodic flows. Let us choose vertices $z_1 \in \omega_1$ and $z_2 \in \omega_2$, $z_1, z_2 \in H$. These vertices are equivalent recurrent ones, which implies the existence of periodic paths ω_0 passing through z_1 and z_2. Consider a path ω of the following form. We start at z_1, make N_1 passes along ω_1, then come to z_2 along ω_0, make N_2 passes along ω_2, and finally return to z_1 along ω_0. Thus, $\omega = N_1\omega_1 + N_2\omega_2 + \omega_0$.

The period of this path is $N = p_1N_1 + p_2N_2 + q$, where p_1, p_2, and q are periods of $\omega_1, \omega_2,$ *and* ω_0 , respectively. According to Proposition 5.4, this periodic path ω generates a flow of the form:

$$m_{ij} = \frac{k(1)_{ij}N_1}{N} + \frac{k(2)_{ij}N_2}{N} + \frac{k(0)_{ij}}{N}, \tag{5.8}$$

where $k(1)_{ij}, k(2)_{ij},$ *and* $k(0)_{ij}$ there are the numbers of passes $\omega_1, \omega_2,$ *and* ω_0 through the edge $i \to j$, respectively. Since $k(0)_{ij}$ is fixed,

$$\frac{k(0)_{ij}}{N} \to 0, \quad \text{при} \quad N \to \infty.$$

We have

$$\frac{k(1)_{ij}N_1}{N} + \frac{k(2)_{ij}N_2}{N} = \frac{k(1)_{ij}p_1N_1}{p_1N} + \frac{k(2)_{ij}p_2N_2}{p_2N} = m(1)_{ij}\frac{p_1N_1}{N} + m(2)_{ij}\frac{p_2N_2}{N}.$$

Choose the integers V and W in such a way that $V, W \to \infty$, and $V/_W \to \alpha$. Since $0 < \alpha < 1$, we have $W > V$. Substituting $N_1 = p_2 V$ and $N_2 = p_1(W - V)$, we obtain

$$N = p_1 p_2 V + p_1 p_2 (W - V) + q = p_1 p_2 W + q,$$

$$\frac{p_1 N_1}{N} = \frac{p_1 p_2 V}{p_1 p_2 W + q} = \frac{V}{W}\left(\frac{p_1 p_2}{p_1 p_2 + q/_W}\right) \to \alpha,$$

$$\frac{p_2 N_2}{N} = \frac{p_1 p_2 (W - V)}{p_1 p_2 W + q} \to 1 - \alpha = \beta,$$

$$m_{ij} = \frac{k(1)_{ij} N_1}{N} + \frac{k(2)_{ij} N_2}{N} + \frac{k(0)_{ij}}{N} \to \alpha m(1)_{ij} + \beta m(2)_{ij}.$$

Thus, the sum of periodic flows $\alpha m(1) + \beta m(2)$ is approximated by constructed periodic flows (5.8). From this, along with Theorem 5.1, we can conclude that every flow from $\mathcal{M}(H)$ can be approximated using periodic flows. The proof has been completed.

5.3 Approximation of Invariant Measures

Consider a homeomorphism $f : M \to$, a measurable partition C, and the symbolic image G generated by this partition. Let us study the space of flows $\mathcal{M}(G)$ under successive subdivision of covering. Let the partition C be subdivided into cells such that each cell $M(i)$ is split into smaller cells $M(i1), M(i2), \cdots$ such that $M(i) = \bigcup_k M(ik)$. Thus, a new partition NC and a new symbolic image NG with vertex indices $\{(ik)\}$ are obtained. The natural map $s : NG \to G$ has a very simple form $s(ik) = i$. This mapping allows us to transfer any flow on NG onto the flow specified on G:

$$s^* : \mathcal{M}(NQ) \to \mathcal{M}(G).$$

According to Propositions 5.2 and 5.3, the mapping $s^* : \mathcal{M}(NG) \to \mathcal{M}(G)$ is continuous and linear, and the image $s^*(\mathcal{M}(NG))$ is a convex compact set in $\mathcal{M}(G)$. It is clear that $s^*(\mathcal{M}(NQ)) \neq \mathcal{M}(G)$.

Definition 5.2 *Two flows $m \in \mathcal{M}(NG)$ and $m^* \in \mathcal{M}(G)$ are called consistent if $s^*(m) = m^*$.*

Consider an invariant measure $\mu \in \mathcal{M}(f)$ and a symbolic image G with respect to a cover. Assume that the cells of this covering are polyhedra that intersect on the boundary disks. Let us construct a partition of these disks by assigning them to one of their adjacent cells. In this way, the mapping $h(x) = \{i : x \in M^*(i)\}$ is uniquely defined. Then, this measure G generates a flow over this symbolic image which is described in Sect. 5.1. The flow is determined by a formula:

$$m = h^*(\mu) = \{m_{ij} = \mu(M^*(i) \cap f^{-1}(M^*(j)))\}.$$

Now consider sequential subdivisions $\mathrm{C}_1,\ \mathrm{C}_2,\ \mathrm{C}_3,\ \cdots$ such that the maximum sell diameter $\mathrm{d}_1,\ \mathrm{d}_2,\ \mathrm{d}_3,\ \cdots$ converges to zero. Such a sequence generates a sequence of symbolic images $\mathrm{G}_1, \mathrm{G}_2, \mathrm{G}_3, \cdots$ and mappings $\mathrm{s} :\ \mathrm{G}_\mathrm{k} \to \mathrm{G}_{\mathrm{k}-1}$. If we have a flow m^k on each symbolic image G_k then we get the sequence of flows $\mathrm{m}^\mathrm{k} \in \mathcal{M}(\mathrm{G}_\mathrm{k})$. The flows m^k for different k, generally speaking, are not related.

Let us show that if each flow m^k is obtained from the invariant measure μ as indicated above, then the sequence measure $\mu\{\mathrm{m}^\mathrm{k}\}$ is consistent. In fact, each mapping $\mathrm{s} :\ \mathrm{G}_\mathrm{k} \to\ \mathrm{G}_{\mathrm{k}-1}$ generates the mapping of flows $\mathrm{s}^* : \mathcal{M}(\mathrm{G}_\mathrm{k}) \to \mathcal{M}(\mathrm{G}_{\mathrm{k}-1})$. Thus, the sequence of mappings of symbolic images is defined as follows:

$$\mathrm{G}_1 \overset{\mathrm{s}}{\leftarrow} \mathrm{G}_2 \overset{\mathrm{s}}{\leftarrow} \mathrm{G}_3 \overset{\mathrm{s}}{\leftarrow} \cdots$$

which generates the sequence of flow mappings

$$\mathcal{M}(G_1) \overset{s^*}{\leftarrow} \mathcal{M}(G_2) \overset{s^*}{\leftarrow} \mathcal{M}(G_3) \overset{s^*}{\leftarrow} \cdots$$

The mapping $f : M \to M$ can be treated as infinite graph with vertices $x \in M$ and edges $x \to f(x)$. For any symbolic image G, there is the mapping of the form $h : M \to G$. This mapping is a mapping of directed graphs. We obtain a sequence of the form:

$$G_1 \overset{s}{\leftarrow} G_2 \overset{s}{\leftarrow} G_3 \overset{s}{\leftarrow} \cdots \overset{h}{\leftarrow} \{f : M \to M\}. \tag{5.9}$$

The sequence (5.9) generates a sequence of the form:

$$\mathcal{M}(G_1) \overset{s^*}{\leftarrow} \mathcal{M}(G_2) \overset{s^*}{\leftarrow} \mathcal{M}(G_3) \overset{s^*}{\leftarrow} \cdots \overset{h^*}{\leftarrow} \mathcal{M}(f). \tag{5.10}$$

This means that each invariant measure generates a consistent sequence of flows on symbolic images.

Let's consider the inverse problem. Let us assume that on each symbolic image G_k there is a flow $m^k \in \mathcal{M}(G_k)$, and that these flows are consistent, that is:

$$s^*\left(m^{k+1}\right) = m^k.$$

Using the Lebesgue measure, we construct a measure μ_k on M for each k:

$$\mu_k(A) = \sum_i m_i^k \frac{\nu(A \cap M(i))}{\nu(M(i))}, \tag{5.11}$$

where A is a measurable set, $M(i)$ is the cell of C_k, and ν is a Lebesgue measure on M. Therefore, we get the sequence of measures $\{\mu_k\}$ on the manifold M.

Theorem 5.3 *Consider successive subdivisions of C_k coverings with the maximal cell diameter tending to zero. If m^k is a consistent sequence of flows on the symbolic images G_k, then there exists an f-invariant measure μ such that*

$$\mu = \lim_{k\to\infty} \mu_k,$$

where the convergence is considered in the weak topology.

Proof. Let $m^k = \left\{m_{ij}^k\right\}$ be a consistent sequence of flows on the symbolic images G_k. Let φ be a continuous function on the compact set and let $C_k = \{M(i)\}$ be a partition on M. To each cell $M(i)$ we assign a measure m_i^k of the vertex $i \in G_k$. Let us take a point $x_i \in M(i)$ and construct the integral sum:

$$F_k(\varphi) = \sum_i \varphi(x_i) m_i^k.$$

Let us show that the limit

$$\lim_{k\to\infty} F_k(\varphi) = F(\varphi) \tag{5.12}$$

exists. It is enough to show that $F_k(\varphi)$ is Cauchy sequence. Let C_l, $l > k$, be a subdivision of the covering C_k so that the cells $M(ir) \in C_l, r = 1, 2, \cdots$ form a subdivision of the cell $M(i)$. Since the sequence of flows is consistent, we have $m^k = s^*\left(m^l\right)$ and

$$m_i^k = \sum_r m_{ir}^l, \tag{5.13}$$

where is the measure of the cell (or the vertices). Let us estimate the difference

$$|F_k(\varphi) - F_l(\varphi)| = \left|\sum_i \varphi(x_i) m_i^k - \sum_{ir} \varphi(x_{ir}) m_{ir}^l\right|$$

Taking into account the equality (5.13) and the uniform continuity of the function on the compact set M, we obtain

$$|F_k(\varphi) - F_l(\varphi)| = \left|\sum_{ir} (\varphi(x_i) - \varphi(x_{ir}))m_{ir}^l\right| \leq \sum_{ir} |\varphi(x_i) - \varphi(x_{ir})|m_{ir}^l \leq$$

$$\max_{|x-y|\leq d_k} |\varphi(x) - \varphi(y)| \sum_{ir} m_{ir}^l = \alpha(d_k),$$

where $\alpha(d)$ is the modulus of continuity of the function φ and d_k is the maximum diameter of C_k. Since $\alpha(d) \to 0$ as $d \to 0$, the sequence $F_k(\varphi)$ is a Cauchy sequence and therefore the limit (5.12) exists.

In the same way, we can show that this limit does not dependent on the choice of $x_i \in M(i)$. Thus, the linear functional $F(\varphi)$ is well defined. It is bounded, because $|F(\varphi)| \leq sup_M |\varphi|$ and positive: $F(\varphi) \geq 0$ for $\varphi > 0$.

According to the Riesz representation theorem (see [1, 3]), there exists a measure such that

$$F(\varphi) = \int_M \varphi d\mu.$$

Since the measure μ_k is defined according to the formula (5.11), the measure μ_k on a cell differs from the Lebesgue measure by a constant factor, and the measure of each cell $\mu_k(M(i)) = m_i^k$ coincides with the measure of the vertex $i \in G_k$. Next, we show that

$$\lim_{k\to\infty} \mu_k = \mu$$

in weak topology. It suffices to show that for any continuous function φ

$$\int_M \varphi \, d\mu_k \to \int_M \varphi d\mu$$

as $k \to \infty$. By the mean value theorem, for each cell $M(i)$, there is a point x_i^* in the closure cell $M(i)$ such that

$$\int_{M(i)} \varphi \, d\mu_k = \varphi(x_i^*)\mu_k(M(i)) = \varphi(x_i^*)m_i^k.$$

Hence

$$\int_M \varphi \, d\mu_k = \sum_i \int_{M(i)} \varphi d\mu_k = \sum_i \varphi(x_i^*)m_i^k.$$

So, it is enough to show that

$$\lim_{k\to\infty}\sum_i \varphi(x_i^*)m_i^k = \lim_{k\to\infty}\sum_i \varphi(x_i)m_i^k,$$

where $|x_i^* - x_i| \le d_k$. This can be proved in the same way as above by using the modulus of continuity of the functions φ.

It is known [1] that the invariance of a measure μ with respect to f follows from the equality:

$$\int_M \varphi d\mu = \int_M \varphi(f)d\mu,$$

where φ is any continuous function on M. Consider the integral sum $F_k(\varphi) = \sum_i \varphi(x_i)m_i^k$, where

$$m_i^k = \sum_j m_{ij}^k = \sum_r m_{ri}^k.$$

We have

$$F_k(\varphi) = \sum_i \varphi(x_i)\sum_r m_{ri}^k = \sum_{ir}\varphi(x_i)m_{ri}^k.$$

We replace the point x_i in each term $\varphi(x_i)m_{ri}^k$ by a point x_{ri} so that $x_{ri} \in f(M(r)) \cap M(i)$, and we get

$$F_k(\varphi) = \sum_{ir}\varphi(x_{ri})m_{ri}^k + \sum_{ri}(\varphi(x_i) - \varphi(x_{ri}))m_{ri}^k = \sum_{ir}\varphi(x_{ri})m_{ri}^k + \varepsilon^*,$$

where ε^* is estimated through the modulus of continuity of the function φ and $\varepsilon^*(d_k) \to 0$ as $k \to \infty$. For each point $x_{ri} \in f(M(r)) \cap M(i)$ there is a point $z_{ri} \in M(r) \cap f^{-1}(M(i))$ such that, $f(z_{ri}) = x_{ri}$, that is, $z_{ri} = f^{-1}(x_{ri})$. Therefore, we have

$$F_k(\varphi) = \sum_{ir}\varphi(f(z_{ri}))m_{ri}^k + \varepsilon^*,$$

where all the points z_{ri}, $i = 1, i = 2, \cdots$, lie in the cell $M(r)$. Let us replace these points with a single point $z_r \in M(r)$. We obtain the equalities

$$F_k(\varphi) = \sum_{ir}\varphi(f(z_r))m_{ri}^k + \sum_{ir}(\varphi(f(z_{ri})) - \varphi(f(z_r)))m_{ri}^k + \varepsilon^* =$$

$$\sum_r \varphi(f(z_r))\sum_i m_{ri}^k + \varepsilon^{**} + \varepsilon^* = \sum_r \varphi(f(z_r))m_r^k = F_k(\varphi(f)) + \varepsilon^{**} + \varepsilon^*,$$

where ε^{**} is estimated through the modulus of continuity of $\varphi, \varphi(f)$, and $\varepsilon^{**}(d_k) \to 0$ as $k \to \infty$. Passing to the limit as $k \to \infty$, we obtain the equality

$$F(\varphi) = F(\varphi(f)),$$

which says that the measure μ is f-invariant, as required. The proof of the theorem is complete.

As stated previously, each invariant measure μ generates a sequence of flows m^k on symbolic images for any subdivision sequence $C_1, C_2, C_3, \cdots$. The sequence m^k is consistent. Theorem 5.3 guarantees that the converse is true, namely, that each consistent flow sequence m^k on the symbolic image G_k generates an invariant measure.

Corollary Any invariant measure μ can be obtained using the method described in Theorem 5.3. Thus, an invariant measure and a consistent sequence of flows are interconvertible.

5.4 Computation of Invariant Measure

Consider successive subdivisions of coverings C_k with the maximal cell diameter $d_k \to 0$ and a consistent sequence of flows m^k on the symbolic images G_k. The flows m^k generate a sequence of measures μ_k, constructed according to the formula (5.11). According to Theorem 5.3, this sequence μ_k converges to an invariant measure μ in the weak topology. Let us now show that this measure can be computed directly using the flows m^k, without using the weak topology.

The cells of the covering $\{C_k\}$ are polyhedra that intersect along boundary disks. We construct a measurable partition by assigning these boundary disks to one of the neighboring cells. Thus, we have a sequence of partitions $C_k^* = \{M_k^*(i)\}$ such that each C_{k+1}^* is a subdivision of C_k^* and there is a mapping $s : G_{k+1} \to G_k$ of directed graphs. This means that each cell $M_k^*(i)$ is a union of cells $M_{k+1}^*(j)$ such j that $s(j) = i$ or $j \in J = s^{-1}(i)$. Since the cells $M_{k+1}^*(j)$ do not intersect, each cell $M_k^*(i)$ is the sum of $M_{k+1}^*(j), j \in s^{-1}(i)$, that is:

$$M_k^*(i) = \sum_{j \in s^{-1}(i)} M_{k+1}^*(j).$$

Since measures μ_k are constructed according to (5.11), the μ_k-measure of a cell $M_k^*(i)$ coincides with the measure of its vertex i for the flow m^k

$$\mu_k\big(M_k^*(i)\big) = m_i^k.$$

Because the flows m^k and m^{k+1} are consistent, we get the equality

$$m_i^k = \sum_{j \in J} m_j^{k+1}, \tag{5.14}$$

which describes the connection between flows m^k and m^{k+1}. Therefore, we obtain equalities for all

$$\mu_k\big(M_k^*(i)\big) = m_i^k = \sum_{j \in J} m_j^{k+1} = \sum_{j \in J} \mu_{k+1}(M_{k+1}^*(j)) = \mu_{k+1}(M_k^*(i)).$$

Subsequently we obtain the equalities for any $t > k$:

$$\mu_k\big(M_k^*(i)\big) = \mu_t(M_k^*(i)).$$

This means that values of measures μ_t on cells $M_k^*(i)$ do not change. Passing the limit as $t \to \infty$ we obtain equality for μ-measure

$$\mu\big(M_k^*(i)\big) = \mu_k\big(M_k^*(i)\big).$$

Thus, we get connection between flows m^k, invariant measure μ and equalities

$$\mu\big(M_k^*(i)\big) = m_i^k \tag{5.15}$$

for any i and k. Our goal is to prove uniqueness of μ-measure for any Borel set. It's enough to find measure of a closed set. Measure of open set is calculated through measure of closed sets. Therefore, measure is defined for all Borel sets, see [3].

Theorem 5.4 Let C_k be successive subdivisions of a covering with the maximal cell diameter tending to zero. If $\{m^k \in \mathcal{M}(G_k)\}$ is a consistent sequence of flows, A is a closed set and μ is the invariant measure constructed in Theorem 5.3, then

$$\mu(A) = \lim_{k \to \infty} \sum_{i \in I} m_i^k, \tag{5.16}$$

where $I = \{i : M_k^*(i) \cap A \neq \emptyset\}$.

Proof The value of $\mu(A)$ is found based on the equalities (5.15). Let us construct a covering of the set A by the cells of C_k^*

$$P_k(A) = \{\cup M_k^*(i), i \in I : M_k^*(i) \cap A \neq \emptyset\}.$$

Let's show that

$$P_k(A) \supset P_{k+1}(A),$$

that is, the sequence $P_k(A)$ is nested. Let us denote $I_k = \{i : M_k^*(i) \cap A \neq \emptyset\}$. The equality $s(j) = i$ means that the cell $M_{k+1}^*(j)$ is included in the subdivision of $M_k^*(i)$. If $M_{k+1}^*(j) \cap A \neq \emptyset$ and $s(j) = i$, then

$$M_k^*(i) \cap A \supset M_{k+1}^*(j) \cap A \neq \emptyset.$$

Therefore, $s(I_{k+1}) \subset I_k$ and

$$P_{k+1} = \{\cup M_{k+1}^*(j), j \in I_{k+1}\} \subset \{\cup M_k^*(i), i \in I_k\} = P_k.$$

We have to prove

$$A = \lim_{k\to\infty} P_k(A) = \bigcap_k P_k(A).$$

By construction, the set A lies in each $P_k(A)$. Consequently, the inclusion $A \subset \bigcap_k P_k(A)$ takes place. Let's prove the converse inclusion by contradiction. Suppose there exists a point $x \in \bigcap_k P_k(A)$ that does not lie A. Since A is a closed set, its distance $\rho(x, A) = r > 0$. This means a cell $M_k^*(i)$ of diameter $d_k < r$ containing the point x cannot intersect A. Therefore, x does not lie at intersection $\bigcap_k P_k(A)$. The resulting contradiction leads to equality

$$\bigcap_k P_k(A) = A.$$

Since the cells $M_k^*(i)$ do not intersect for different values k, we obtain equalities

$$\mu(A) = \lim_{k\to\infty} \mu(P_k(A)) = \lim_{k\to\infty} \sum_{i\in I} \mu(M_k^*(i)) = \lim_{k\to\infty} \sum_{i\in I} \mu\big(M_k^*(i)\big) = \lim_{k\to\infty} \sum_{i\in I} m_i^k,$$

where $I = \{i : M_k^*(i) \cap A \neq \emptyset\}$. The limits above exist due to the properties of monotonicity and additivity of measure. The proof is complete.

Corollary If is $B \subset M$ an open set, then

$$\mu(B) = \mathit{lim}_{k\to\infty} \sum_{j\in J} m_j^k,$$

where $J = \{j : M_k^*(j) \subset B\}$.

Proof. The set $A = M \backslash B$ is closed and

$$\mu(B) = \lim_{k\to\infty} \sum_{j\in J} m_j^k,$$

where $I = \{i : M_k^*(i) \cap A \neq \emptyset\}$. If a cell $M_k^*(i)$ does not intersect the set $A = M \backslash B$, then $M_k^*(i)$ lies in B. Therefore, the set

$$P_k(A) = \underset{i\in I}{\cup} M_k^*(i)$$

is complementary to the set

$$P_k(B) = \underset{j\in J}{\cup} M_k^*(j)$$

on the manifold M. From the decreasing sequence $P_k(A) \to A$ it follows that $P_k(B)$ the monotonically increases and converges to $B = M \backslash A$. We have

$$\mu(B) = \lim_{k\to\infty} \mu(\underset{j\in J}{\cup} M_k^*(j)) = \lim_{k\to\infty} \sum_{j\in J} m_j^k.$$

The proof is over.

5.5 Sequence of Non-consistent Flows

We will now study the inconsistent flow sequences. Consider the sequence of symbolic images $G_1, G_2, \cdots, G_t, \cdots$ of a homeomorphism with respect to the sequence of divisions $C_1, C, \cdots, C_t, \cdots$ with $d_t \to 0$ as $t \to \infty$. Fix a flow m^t on each symbolic image G_t. The flows m^t are, generally speaking, inconsistent. Using the flows, we construct the sequence of measures on by the formula (5.11). On each symbolic image G_τ, we define the sequence of flows m_τ^k as a projection of flows $m^{\tau+k}$ by means of the mapping $s^* : \mathcal{M}(G_{\tau+k}) \to \mathcal{M}(G_\tau)$. For the space of flow on G, we introduce a distance $\rho\left(m^1, m^2\right) = sum_{ij}\left|m_{ij}^1 - m_{ij}^2\right|$, where m_{ij}^* is the measure of edge $i \to j$. The measure of cell $M(i)$ is equal to $m_i^* = \sum_j m_{ij}^*$.

Definition 5.3 *A sequence of flows m^t is said to converge if the sequence of the projections m_τ^k converges in the distance ρ on each G_τ as $k \to \infty$.*

Theorem 5.5 *If a sequence of flows m^t converges, then a sequence of measures μ_t constructed by formula (5.1) converges to invariant measure in weak topology.*

Proof. On each symbol image G_τ, we fix a flow $m_\tau^* = \lim_{k\to\infty} m_\tau^k$. By construction, flows m_τ^* are consistent, i.e., $s_{\tau+1}^*\left(m_{\tau+1}^*\right) = m_\tau^*$. According to Theorem 5.4, consistent sequence m_τ^* generates a sequence of converges to invariant measure in weak topology.

Now we show that original sequence μ_t converges to μ in weak topology as well. Let $t = \tau + k$ and let G_τ be a symbol image with respect to covering C_τ, let $M(i)$ be cells of C_τ. Let φ be continuous function on M. We have

$$\int \varphi d\mu_t - \int \varphi d\mu = \sum_i \varphi(x_i)m_i^t - \sum_i \varphi\left(x_i^*\right)m_i^*,$$

where $m_i^t = \mu_t(M(i)), m_i^* = \mu(M(i))$, and the points x_i and x_i^* lie in the closure $\overline{M(i)}$ and are determined by the mean value theorem applied to each cell. By assumption we have $\sum_i \left|m_i^t - m_i^*\right| =$

$$\sum_i \left|\mu_t(M(i)) - \mu^*(M(i))\right| = \sum_i \left|m_\tau^k(M(i)) - m_\tau^*(M(i))\right| \leq \rho\left(m_\tau^k, m_\tau^*\right) \to 0$$

as $k \to \infty$. Hence

$$|\int \varphi d\mu_t - \int \varphi d\mu| = |\sum_i [\varphi(x_i)(m_i^t - m_i^*) + \left(\varphi(x_i) - \varphi\left(x_i^*\right)\right)m_i^*]| \leq$$
$$||\varphi|| \sum_i \left|m_i^t - m_i^*\right| + \alpha(d_\tau) \sum_i m_i^* \leq ||\varphi|| \rho\left(m_\tau^k, m_\tau^*\right) + \alpha(d_\tau),$$

where $||\varphi|| = sup_M |\varphi(x)|$, d_τ is the maximal diameter of C_τ, and α(d) is the modulus of continuity of the function φ. To prove that $\mu_t \to \mu$ in the weak topology it is enough to show that for a given function φ and any $\varepsilon > 0$ there exists a number t_0 such that

$$\left|\int \varphi d\mu_t - \int \varphi d\mu\right| < \varepsilon$$

for $t > t_0$. For any given number $\varepsilon/2$ and the function φ we fix τ and $d_\tau > 0$ such that $\alpha(d_\tau) < \varepsilon/2$. For fixed values of $\tau, \varepsilon/2$ and $||\varphi||$, we find k such that $\rho\left(m_\tau^k, m_\tau^*\right) < \varepsilon/(2||\varphi||)$, if $1 \geq ||\varphi||$ or $\rho\left(m_\tau^k, m_\tau^*\right) < \varepsilon/2$ if $||\varphi|| < 1$. Set $t_0 = \tau + k$. Then for $t > t_0$ we have

$$\left|\int \varphi d\mu_t - \int \varphi d\mu\right| \leq ||\varphi|| \rho\left(m_\tau^k, m_\tau^*\right) + \alpha(d_\tau) < \frac{\varepsilon}{2} + \frac{\varepsilon}{2} = \varepsilon.$$

Thus, $\int \varphi d\mu_t \to \int \varphi d\mu$ as $t \to \infty$, and therefore the sequence of measures μ_t converges to μ in weak topology. The proof is complete.

If the sequence m^t is consistent, then $s^*\left(m^{t+1}\right) = m^t, m_\tau^k = m^\tau$ and the sequence of projections m_τ^k converges to m^τ. So, Theorem 5.5 is a generalization of Theorem 5.3. These theorems cannot be applied to sequences of general form. The following theorem describes the properties of arbitrary sequences of flows.

Theorem 5.6 *Suppose that the sequence of symbolic images $\{G_t\}$ of the homeomorphism f and the sequence of flows $\left\{m^t\right\}$ are fixed and that the diameter of coverings $d_t \to 0$ as $t \to \infty$. Then*

1. there exists a subsequence indexed *by* $t_k \to \infty$ such that the measures μ_{t_k} constructed via the formula *(5.11)* converge in the weak topology to an invariant measure μ;
2. if some subsequence μ_{t_l} converges in weak topologies to the measure μ^*, *then* it is invariant.

Proof. Let's consider one symbolic image G_1. The set of flows forms a polyhedron $\mathcal{M}_1$, which is the convex hull of simple flows. Each flow $m = \left\{m_{ij}\right\}$ is represented by a point in R^N, where N is the number of vertices of $\mathcal{M}_1$. The distance $\rho\left(m^1, m^2\right)$ is defined on $\mathcal{M}_1$. Using the natural map $s^* : \mathcal{M}(G_t) \to \mathcal{M}(G_1) = \mathcal{M}_1$, we transfer flows m^t onto the graph G_1 and denote the transformed flow m^t; that is, $m_1^t = s^*\left(m^t\right)$. So, on compact $\mathcal{M}_1$, there is sequence $\left\{m_1^t\right\}$, from which a convergent subsequence can be taken. Without loss of generality, let's assume that original sequence is convergent. Let $m^*{}_1 = \lim_{t\to\infty} m^t{}_1$. Next, consider symbolic image G_2 for which flow m_2^* is constructed as limit of certain subsequence projections $s^*\left(m^t\right)$ onto $\mathcal{M}_2$. Following this procedure, we construct flow m_t^* on each symbolic G_t. If we take diagonal sequence m^{t_k}, then sequence $s^*\left(m^{t_k}\right)$ converges to m_t^* on each symbolic image G_t. According to Theorem 5.5, sequence of measures μ_{t_k} converges to invariant measure μ in weak topology, thus first assertion of theorem is proven.

To proof of the second statement, suppose that there is a subsequence of measures converging to the measure μ in the weak topology. Without loss of generality, we can assume that the original sequence μ_t converges to μ. By the fist assertion, we can take a subsequence μ_t that converges to an invariant measure μ^*. Consequently, $\mu^* = \mu$ by uniqueness of the limit, and the measure is invariant. The theorem is proven.

The results we have obtained are applicable to sequence of flows with $d_k \to 0$. In practice, it is desirable to have a result concerning an individual flow on a symbolic image for a small positive diameter d.

Theorem 5.7 *For any neighborhood U (in the weak topology) of the set $\mathcal{M}(f)$ there exists a positive d_0 such that for any covering C with maximum diameter cells satisfying $d < d_0$ and for any flow m on the symbolic image G constructed with respect to C, the measure μ (constructed via m and (5.11)) lies in the neighborhood U.*

Proof. Suppose, on the contrary, there exists a neighborhood U of the set $\mathcal{M}(f)$ such that for any cover C_k with maximum diameter d_k on the symbolic image G_k

there is a flow m_k for which the measure μ_k does not lie in the neighborhood U. We may assume this $d_k \to 0$. According to Theorem 5.6, there is a subsequence $\{m_{k_p}\}$ such that the sequence of measures μ_{k_p} constructed according to (5.11) converges to an invariant measure $\mu \in \mathcal{M}(f)$. This means that for some k_{p_0} the subsequence μ_{k_p} with $p > p_0$ is in U, which contradicts the assumption.

Theorem 5.7 guarantees that any measure constructed by means of symbolic image is an approximation of some invariant measure provided the cell diameter is sufficiently small. In practice, we can only construct a finite number of symbolic images. The obtained result provides a basis for practical computations. Moreover, this theorem allows us to view the set of all flows $\mathcal{M}(G)$ on a symbolic image as approximations of the set $\mathcal{M}(f)$ of invariant measures provided the diameter of the cover is sufficiently small.

5.6 Approximation by δ-Measures

The approximation of an invariant measure is constructed according to the following schema. We take a covering $C = \{M(i)\}$, symbolic image $C = \{M(i)\}$ and a flow $m = \{m_{ij}\}$ are constructed on it. Then we determine the desired measure μ on M by setting

$$\mu(A) = \sum_i m_i \frac{\nu(A \cap M(i))}{\nu(M(i))}, \tag{5.17}$$

where $m_i = \sum_j m_{ij} = \sum_k m_{ki}$, ν is the Lebesgue measure on M. In this case, the measure of the cell $M(i)$ coincides with the measure of the vertex i:

$$\mu(M(i)) = m_i.$$

The measure is an approximation to an invariant measure in the sense that μ converges in a weak topology to an invariant measure as the maximum cell diameter tends to zero. According to the equality (5.17), we extend the measure to the subsets of the cell $M(i)$ by using the Lebesgue measure ν. The question arises: how important is it to use the Lebesgue measure when constructing the measure μ? If $\varphi(x)$ is a continuous function then, by the mean value theorem, for each cell $M^*(i)$ there exists a (mean) point x_i such that

$$\int_{M^*(i)} \varphi d\mu = \varphi(x_i)\mu\big(M^*(i)\big) = \varphi(x_i)m_i.$$

All that is known about the point x_i is that it lies in the closed cell $M(i)$. If we use a different measure ν (other than the Lebesgue measure), the point x_i may change while remaining in the cell $M(i)$.

Proposition 5.7 *Let $m = \{m_{ij}\}$ be the flow on the symbolic image with respect to the covering $C = \{M(i)\}$. The cells $M(i)$ are polyhedrons intersecting along the boundary disks, a measurable partition $C^* = \{M^*(i)\}$ is obtained by assigning boundary disks to one of the neighboring cells. The measure μ be constructed according to the formula (5.17). Let us assume that the measure μ^* be such that*

(1) *cell measure $\mu^*(M(i)) = m_i = \sum_j m_{ij}$,*
(2) *for any continuous function φ, the equalities*

$$\int\limits_{M^*(i)} \varphi d\mu^* = \varphi(x_i^*) m_i, \; x_i^* \in M(i)$$

valid for every $M(i)$.

Then

$$\left| \int\limits_M \varphi d\mu - \int\limits_M \varphi d\mu^* \right| \le \alpha(d),$$

where d is the diameter of the covering and $\alpha(.)$ *is the modulus of continuity of the function $\varphi(x)$.*

Proof. For the partition $C^* = \{M^*(i)\}$, we have

$$\int\limits_M \varphi d\mu = \sum_i \int\limits_{M^*(i)} \varphi d\mu = \sum_i \varphi(x_i)\mu(M^*(i)) = \sum_i \varphi(x_i) m_i$$

and

$$\int\limits_M \varphi d\mu^* = \sum_i \int\limits_{M^*(i)} \varphi d\mu^* = \sum_i \varphi(x_i^*)\mu^*(M^*(i)) = \sum_i \varphi(x_i^*) m_i,$$

where the points x_i and x_i^* lie in the same cell $M(i)$. Then

$$\left| \int\limits_M \varphi d\mu - \int\limits_M \varphi d\mu^* \right| \le \sum_i |\varphi(x_i) - \varphi(x_i^*)| m_i \le \alpha(d).$$

The proposition is proved.

A consequence of this proposition is Theorem 5.8.

Theorem 5.8 *Consider the sequence of symbolic images G_t for coverings with diameters $d_t \to 0$ and a sequence of flows m_t on G_t. Suppose that there are two sequences of measures on M μ_t and μ_t^*, where μ_t is constructed using the formula (5.17), and the measure μ_t^* satisfies the conditions of Proposition 5.7 for each t. Then μ_t and μ_t^* converge to a common limit in the weak topology.*

This implies that all previous results on approximation of invariant measures remain valid for measures described in Proposition 5.7.

Now we construct an approximation to the invariant measure by a measure that is concentrated on a finite set of points. Let $m = \{m_{ij}\}$ be a flow on the symbolic image. Let $\delta(x)$ be the Dirac measure (function) concentrated at point x; that is, for any continuous function φ

$$\int_M \varphi d\delta(x) = \varphi(x).$$

In each cell $M^*(i)$ we select a point x_i. Let us consider the measure of the form:

$$\mu^* = \sum_i m_i \delta(x_i). \tag{5.18}$$

We call the constructed measure the discrete measure concentrated at the points $\{x_i\}$. For the discrete measure μ^*, we have

$$\int_M \varphi d\mu^* = \sum_i \varphi(x_i) m_i, \quad x_i \in M(i).$$

The measure of the cell is $\mu^*(M(i)) = m_i$. Thus, all the conditions of Proposition 5.7 are met, and therefore the following theorem follows from Theorem 5.8.

Theorem 5.9 *Let the sequence of flows $\{m^t\}$ be defined on the sequence of symbolic images G_t of the homeomorphism f, and let the maximum diameters of the covering $d_t \to 0$ as $t \to \infty$. Then*

(1) *there exists a subsequence of discrete measures μ_{t_k} constructed according to formula (5.18) that converging to an invariant measure μ in the weak topology;*
(2) *if some subsequence of discrete measures μ_{t_l} converges in the weak topology to a measure μ^*, then it is invariant for f.*

Theorem 5.10 For any neighborhood (in the weak topology) U of the set $\mathcal{M}(f)$ there exists a number d_0 such that for any covering C with diameter $d < d_0$ and any flow m on the symbolic image G constructed to C, a discrete measure μ constructed according to (5.18) lies in the neighborhood U.

5.7 Balance Method

We looked at constructing a flow in a graph through a set of periodic paths. Now, we will consider another method which allows us to obtain an invariant flow from a non-invariant distribution.

Let $P = (p_{ij})$ be an arbitrary distribution of non-negative values on the edges of a graph G. The set of these edges $G(P^+) = \{i \to j : p_{ij} > 0\}$ is called the support of P. Let's construct a flow over the support. From Theorem 5.1, it follows that for a flow to exist over $G(P^+)$, the support must necessarily contain a cycle. For simplicity, we can assume the graph consists of one strongly connected component.

A graph G has n vertices. A matrix $x = (x_{ij})$ is a flow if the following conditions hold:

(1) $x_{ij} \geq 0$,
(2) $\sum_{ij} x_{ij} = 1$,
(3) $\sum_j x_{ij} = \sum_k x_{ki}$ for $i = 1, 2, \cdots, n$.

Moreover, we require the support to x be in $G(P^+)$; that is, $x_{ij} = 0$ if $p_{ij} = 0$. A similar computational task arises in the theory of convex programming. Our design is based on article [8], where the method we need has been justified. The task of computing a flow in a graph can be considered as a special case of a transport problem. Leningrad architect G.V. Sheleikovskii solved a similar transport problem in the 1940s using the balancing method. Starting with an arbitrary distribution, he recomputed the distribution in a sequential manner such that, at each step, only one equality was satisfied while other equalities were ignored. By repeating this process cyclically, he obtained a sequence of solutions that quickly converged to the desired solution. We will solve the problem in the same manner.

The matrix $x = (x_{ij})$ is defined by n^2 numbers, and the flow imposes $n + 1$ conditions (equations) on these numbers. Consequently, we have some freedom to choose, which we use to solve an extra extreme problem.

Theorem 5.11 *Let $P = (p_{ij})$ be an arbitrary non-negative distribution on G such that the support $G(P^+)$ contains a periodic path. Then there exists an algorithm that constructs a flow $Q = (q_{ij})$ with $G(Q^+) \subset G(P^+)$, which maximizes the function*

$$g(x) = \sum_{ij} x_{ij} ln \frac{p_{ij}}{x_{ij}} = \sum_{ij} x_{ij} \ln p_{ij} - \sum_{ij} x_{ij} \ln x_{ij}$$

in the space of flows on $G(P^+)$.

Remark Here, as usual, we set $0 ln 0 = 0$ by continuity.

Proof Let d be the number of edges in the graph $G(P^+)$, with $d \leq n^2$. If the elements with $p_{ij} > 0$ are arranged sequentially $(ij) \to k = 1, 2, \cdots, d$ (for instance, row by row from top to bottom), then x can be considered as a point in space $S = \{x \in R^d : x_k = x_{ij} > 0\}$. Set

$$B_0 = \{x \in R^d : x_{ij} > 0, \ \sum_{ij} x_{ij} = 1\},$$

$$B_i = \{x \in R^d : x_{ij} > 0, \ \sum_j x_{ij} = \sum_k x_{ki}\}$$

for $i = 1, 2, \cdots, n$. The intersection $D = B_0 \cap_i B_i$ is a compact set lying in the closure $\overline{S}$, so the function g reaches its maximum at D. Our goal is to find this point. In [8], the following problem of convex programming was solved.

Let $f(x)$ be a strictly convex, continuously differentiable function on a convex set $S \subset R^d$ and continuous on its closure $\overline{S}$. It is required to minimize the function f under the linear restrictions:

$$Ax = b, x \in \overline{S},$$

where $b \in R^m$, A is a matrix with r columns and m rows. Let's apply the results from [8] to the function $f = -g$, which is a strictly convex. Our linear restrictions are given by the $n + 1$ equation, which normalizes

$$\sum_{ij} x_{ij} = 1, \tag{5.19}$$

and the next equations describe invariance (Kirchhoff's law)

$$\sum_j x_{ij} - \sum_k x_{ki} = 0, \quad i = 1, 2, \cdots, n. \tag{5.20}$$

The necessary condition for the existence of a solution to an optimization problem is the existence of solutions to the system (5.19–5.20). The existence of periodic paths guarantees the existence of flows, which are necessary solutions. Consider the following iterative process:

1. Take an arbitrary point (matrix) $x^0 \in S$.
2. If a point x^t is known, choose the p_t-th equation and find a point x^{t+1} that satisfies this equation, but which may not satisfy the others. The method for computing this point will be described later.

In an iterative loop, you have to take all $n+1$ equations. The sequence $\{x^t\}$ obtained in this way is called a relaxation sequence, and the control sequence $\{p_t\}$ is called a relaxation control. It is possible to take relaxation control in a usual cyclic order: first solve the normalizing equation and then solve successive invariance equations.

It has been shown in [8] that for x^t to solve the extremum problem the points x^t and x^{t+1} must satisfy the following systems of equations:

$$gradf\left(x^{t+1}\right) = gradf\left(x^{t}\right) + \lambda A_p, \tag{5.21}$$

$$\left(A_p, x^{t+1}\right) = b_p, \tag{5.22}$$

where *gradf* is the gradient of f, λ is an unknown parameter, (*,*) denotes the inner product, A_p is the row of coefficients of the left-hand side of the p-th equation, and b_p is the right-hand side of the same equation. In other words, x^{t+1} is a solution of the p-th equation (5.22). For the function

$$f(x) = -g(x) = \sum\nolimits_{ij} x_{ij} ln\frac{p_{ij}}{x_{ij}}$$

we have

$$(gradf(x))_{ij} = ln\frac{x_{ij}}{p_{ij}} + 1.$$

For the normalizing equation (5.19), we get the system

$$ln\frac{x_{ij}^{t+1}}{p_{ij}} = ln\frac{x_{ij}^{t}}{p_{ij}} + \lambda,$$

$$\sum_{ij} x_{ij}^{t+1} = 1.$$

From the first equation, we obtain the equality

$$x_{ij}^{t+1} = x_{ij}^{t} exp(\lambda),$$

and from the second, we have

$$\sum_{ij} x_{ij}^{t} exp(\lambda) = 1.$$

Hence, we obtain the ordinary transformation of normalization

$$x_{ij}^{t+1} = \frac{x_{ij}^{t}}{\sum_{kl} x_{kl}^{t}}. \tag{5.23}$$

For the i-th equation of the invariance (5.20), we obtain the following system of equalities:

$$ln\frac{x_{ij}^{t+1}}{p_{ij}} = ln\frac{x_{ij}^{t}}{p_{ij}} + \lambda,$$
$$ln\frac{x_{ki}^{t+1}}{p_{ki}} = ln\frac{x_{ki}^{t}}{p_{ki}} - \lambda,$$
$$\sum_{j\neq i} x_{ij}^{t+1} - \sum_{k\neq i} x_{ki}^{t+1} = 0.$$

Then

$$exp\lambda = \left(\frac{\sum_{k\neq i} x_{ki}^t}{\sum_{j\neq i} x_{ij}^t}\right)^{\frac{1}{2}}$$

and we get the transformation in the form:

$$x_{ij}^{t+1} = x_{ij}^t\left(\frac{\sum_{k\neq i} x_{ki}^t}{\sum_{l\neq i} x_{il}^t}\right)^{\frac{1}{2}} \tag{5.24}$$

for $j \neq i$,

$$x_{ki}^{t+1} = x_{ki}^t\left(\frac{\sum_{l\neq i} x_{il}^t}{\sum_{m\neq i} x_{mi}^t}\right)^{\frac{1}{2}} \tag{5.25}$$

for $k \neq i$, and

$$x_{ii}^{t+1} = x_{ii}^t. \tag{5.26}$$

Formulas (5.24), (5.25), and (5.26) describe the transformation of the i-th row and i-th column of the matrix at step t. The other elements do not change. According to [8], the sequence obtained in this way has a limit if the function

$$D(x, y) = \sum_{ij} p_{ij}\big(y_{ij} - x_{ij}\big) + \sum_{ij} x_{ij}ln\frac{x_{ij}}{y_{ij}}$$

is such that $D(x, y_k) \to 0$ when $x \in \overline{S}$, $y_k \in S$, and $y_k \to x$. In our case, this property is easily checked. Moreover, in this case, the limit does not depend on the relaxation control, but only on the initial value of x^0. For the limit value

$$Q = \lim_{t\to\infty} x^t$$

to solve the extreme problem, it is enough to take the initial value as the global minimum point of the function f, that is, set $x_{ij}^0 = p_{ij}exp(-1)$. Taking into account the normalization, we can start the iteration process with

$$x_{ij}^0 = p_{ij}. \tag{5.27}$$

Theorem is proven.

The proof gives the formulas (5.23), (5.24), (5.25), (5.26) for the iteration process of calculating a sequence of distributions converging to the desired flow.

5.8 Application of Balance Method

Let us consider the numerical results of calculating the invariant measure of the modified Ikeda mapping. The invariant measure is constructed as a flow on a symbolic image. The constructed invariant measure maximizes the function $\varphi(x_{ij}) = -\sum_{ij} x_{ij} ln x_{ij}$.

The Ikeda mapping arises in the modeling of optical recording media (crystals) [9, 10] and is of the form:

$$\begin{pmatrix} x \\ y \end{pmatrix} \rightarrow \begin{pmatrix} \mathrm{d} + \mathrm{a}(\mathrm{x}\cos(\tau(\mathrm{x},\ \mathrm{y})) - \mathrm{y}\sin(\tau(\mathrm{x},\ \mathrm{y})) \\ \mathrm{b}(\mathrm{x}\sin(\tau(\mathrm{x},\ \mathrm{y})) + \mathrm{y}\cos(\tau(\mathrm{x},\ \mathrm{y})) \end{pmatrix}, \tag{5.28}$$

$$\tau(\mathrm{x}, \mathrm{y}) = 0.4 - \frac{6}{1 + x^2 + y^2}, \tag{5.29}$$

where $\mathrm{d} > 0,\ 0 < \mathrm{a} < 1, 0 < \mathrm{b} < 1$. If $\mathrm{d} = 2,\ \mathrm{a} = -0.9,\ \mathrm{b} = 0.9$, the mapping contracts and has a global chaotic attractor in the domain $\mathrm{M} = [-10, 10]\times[-10, 10]$. To find it, we use ten subdivisions to construct a sequence of images $\mathrm{G}_1, \ldots, \mathrm{G}_{10}$. In the top of Fig. 5.2, we see the attractor lying in the (x, y)-plane. It consists of 96543 cells of size 0.019×0.019. The invariant flow is created using the balance method with an adjacency matrix $\{\mathrm{p}_{\mathrm{ij}} = \pi_{\mathrm{ij}}\}$ as the initial value.

The balance method leads to a matrix $\{m_{ij}\}$ of size 96543×96543. The resulting distribution is a probability measure on the vertices of the symbolic image, which is an approximation of the invariant measure, according to Theorem 5.7. From Theorem 5.11, it follows that this distribution maximizes the function $\varphi(x_{ij}) = -\sum_{ij} x_{ij} ln x_{ij}$. Therefore, the invariant measure is distributed over the chaotic attractor as much as possible. The distribution of this measure is shown on the lower panel of Fig. 5.2, with each cell's measure represented by a Z-value.

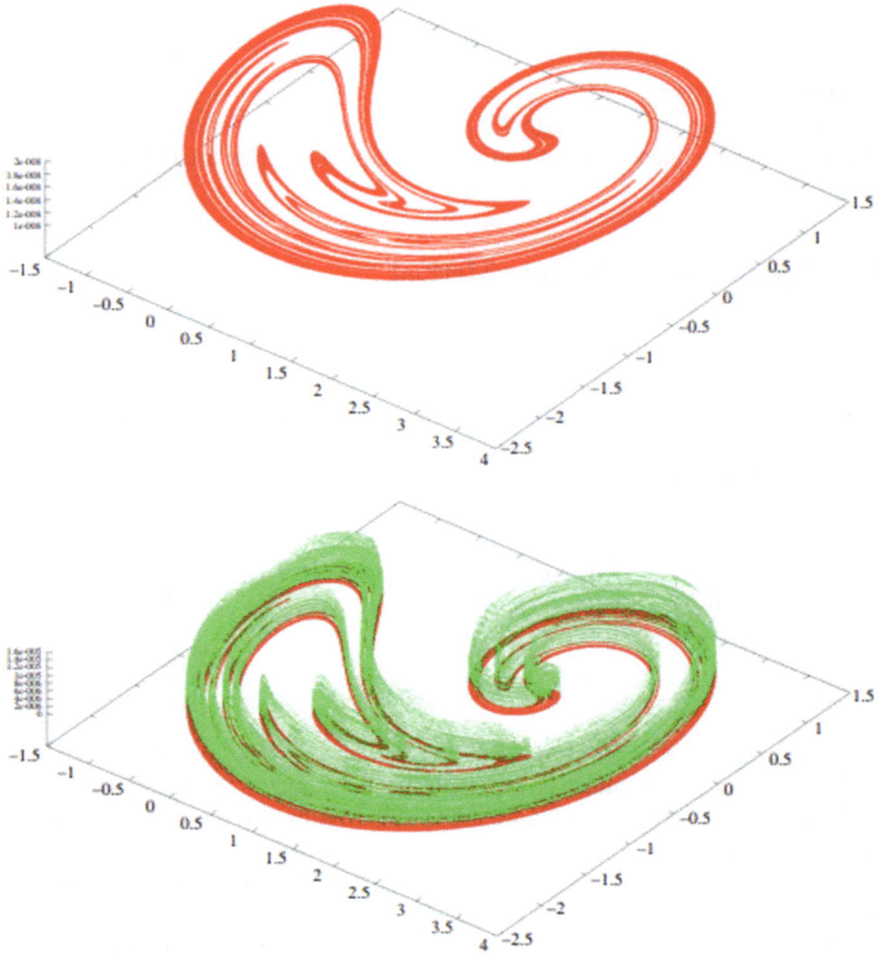

Fig. 5.2 Balance method for Ikeda mapping

References

1. Katok, A., Hasselblatt, B.: Introduction to the Modern Theory of Dynamical Systems. Cambridge University Press (1995)
2. Kryloff, N., Bogoliouboff, N.: La théorie générale de la mesure dans son application a l'etude des systemes dynamiques de la mécanique non linéaire. Ann Math **38**(1), 65–113 (1937)
3. Nemytski, V.V., Stepanov, V.V.: Qualitative Theory of Dynamical Systems. Princeton University Press. Russian original 1949 (1960)
4. Billingsley, P.: Ergodic Theory and Information. Wiley, New York (1965)
5. Ulam, S.: Problems in Modern Mathematics. Interscience (1964)
6. Froyland, G.: On Ulam approximation of the isolated spectrum and eigenfunctions of hyperbolic maps. Discret. Contin. Dynam. Syst. **17**(3), 203–221 (2007)
7. Ding, J., Zhou, A.: Finite approximations of Frobenious-Perron operators. A solution of Ulam's conjecture to multi-dimensional transformation. Physica D **92**(1–2), 61–68 (1992)

8. Bregman, L.M.: Relaxation method for finding a common points of convex sets and its application for convex programming problems. J. Comput. Math. Math. Phys. **7**(3), 630–631 (1967). (in Russian)
9. Ikeda, K.: Multiple-valued stationary state and its instability of the transmitted light by a ring-cavity system. Opt. Comm. **30**, 257–261 (1979)
10. Ikeda, K., Daido, H., Akimoto, O.: Optical turbulence: chaotic behavior of transmitted light from a ring cavity. Phys. Rev. Lett. **45**, 709–712 (1980)

Chapter 6
Ergodic Measures

6.1 Basic Information

Let $C(M)$ be the space of continuous functions on M with norm $|\varphi| = max|\varphi(x)|$. A measure μ on M defines a linear positive functional on $C(M)$:

$$L(\varphi)=\int_M \varphi\, d\mu,\quad L(\varphi)\geq 0,\ \text{if}\,\varphi\geq 0.$$

Riesz's theorem [1] states that each positive linear functional L on $C(M)$ corresponds to some measure μ such that $L(\varphi) = \int_M \varphi d\mu$. Moreover, this correspondence is one-to-one. In particular, measures μ_1 and μ_2 are different if there exists a continuous functional such that

$$\int_M \varphi d\mu_1 \neq \int_M \varphi d\mu_2.$$

Normalized measures correspond to functionals satisfying condition $L(1) = 1$. In the space of measures $\{\mu\}$ (or in the space of linear functionals $\{L\}$), there is a weak topology in which a sequence μ_k converges to μ if and only if, for every continuous function $\varphi \in C(M)$, the sequence $\int_M \varphi d\mu_k$ converges to $\int_M \varphi d\mu$.

The Ergodic Theorem of Birkhoff–Khinchin [1–3]. *Let* (M, μ) *be a space with an invariant probability measure for* f. *Then:*

(1) *for any function* $\varphi \in L^1(M, \mu)$, *there is an average (over time)*

$$\overline{\varphi}(x) = \lim_{n\to\infty} \frac{1}{n}\sum_{k=0}^{n-1} \varphi\big(f^k(x)\big)$$

G. Osipenko, *Symbolic Analysis of Dynamical Systems*, Mathematical Marvels: Texts and Monographs in the Spirit of CR Rao, https://doi.org/10.1007/978-981-95-0713-9_6

for almost all x in measure μ;

(2) $\overline{\varphi}$ *is invariant under* f, *that is,* $\overline{\varphi}(x) = \overline{\varphi}(f(x))$; $\overline{\varphi} \in L^1(M, \mu)$ *and*

$$\int_M \overline{\varphi} d\mu = \int_M \varphi d\mu.$$

Theorem 6.1 [3] *If a homeomorphism* f *has two ergodic measures* μ_1 *and* μ_2, *then they are mutually singular, that is, there is a measurable partition* $M = A + B$ *such that* $\mu_1(A) = 1$ *and* $\mu_2(B) = 1$.

Equivalent formulations of ergodicity [2–4]:

(1) An invariant measure μ is ergodic if and only if the measure of every invariant measurable set is either 1 or 0.
(2) An invariant measure μ is ergodic if its time average equals its space average

$$\lim_{n\to\infty} \frac{1}{n} \sum_{k=0}^{n-1} \varphi\big(f^k(x)\big) = \int_M \varphi d\mu$$

for almost all x with respect to measure μ, where $\varphi \in C^0(M)$ or $\varphi \in L^1(M, \mu)$.

([3]) An invariant measure μ is ergodic if any function $\varphi \in C^0(M)$ or $\varphi \in L^1(M, \mu)$ invariant for f (i.e., $\varphi(x) = \varphi(f(x))$) is constant almost everywhere in measure μ.

Let denote $ext(\mathcal{M})$ the set of extreme points of convex set $\mathcal{M}$. It is known (see [4]) that extreme points in $\mathcal{M}(f)$ are ergodic measures. Thus, we have an equivalent formulation:

(4) An invariant measure μ is ergodic if it is the extreme point of set of invariant measures: $\mu \in ext(\mathcal{M}(f))$.

This definition of ergodicity will be used extensively in future.

6.2 Approximation of Ergodic Measures

The set of invariant measures $\mathcal{M}(f)$ is a convex compact in weak topology whose extreme points are ergodic measures. The convex hull of a set A is the intersection of all convex sets containing it. A convex hull is a convex set. According to the Theorem 1.2.2 from [5], the convex hull of the set A can be defined as the set of all possible convex combinations of a finite number of points from A. The convex hull defined in this way does not have to be a closed set. Let $\mathcal{M}^0$ denote the set of measures, each of which can be expressed as a convex combination of a finite number of ergodic ones.

$$\mathcal{M}^0 = \left\{ \mu = \sum_{k=1}^{m} \alpha_k \mu_k,\ \mu_k \in ext\ \mathcal{M}(f),\ \alpha_k \geq 0,\ \sum_k \alpha_k = 1 \right\}.$$

$\mathcal{M}^0$ is convex and, moreover,

$$\mathcal{M}^0 \subset \mathcal{M}(f).$$

The extreme points of $\mathcal{M}^0$ and $\mathcal{M}(f)$ coincide. According to [1], the closure of $\mathcal{M}^0$ is $\mathcal{M}(f)$:

$$\overline{\mathcal{M}^0} = \mathcal{M}(f).$$

Choquet's theorem states that any point in convex compact set can be represented as an integral sum over its extreme points. We also need the following result about expansion of systems with finite invariant measure into ergodic parts (see Theorem 4.1.12 in [2]).

Theorem 6.2 *For each invariant measure $\mu \in \mathcal{M}(f)$, there is a partition (modulo nullsets) of M into invariant measurable sets $\Omega_\alpha, \alpha \in A$, where A is an abstract measure space, and each Ω_α is equipped with ergodic invariants measure μ_α such that for any integrable function φ,*

$$\int_M \varphi d\mu = \int_A \int_M \varphi d\mu_\alpha d\alpha. \tag{6.1}$$

Let C be a covering of M by polyhedral cells that intersect in boundary disks, and let C^* be the partition obtained from C by assigning each boundary disk to one of adjacent polyhedra. Let G be the symbolic image with respect to C. The set-valued mapping $h : M \to G, h(x) = \{i : x \in M(i)\}$ is single-valued on C^*. This mapping gives rise to a mapping h^* from the set of invariant measures to flows $\mathcal{M}(G)$ such that $m_{ij} = \mu(f(M^*(i)) \cap M^*(j))$,where $\mu \in \mathcal{M}(f)$. The mapping h^* is linear in the sense that it satisfies the equation

$$h^*(\alpha\mu_1 + \beta\mu_2) = \alpha h^*(\mu_1) + \beta h^*(\mu_2),$$

where α and β are constants. Note that $h^* : M(f) \to M(G)$ is not necessarily continuous as a function from $\mathcal{M}(f)$ with weak topology to the space $\mathcal{M}(G)$ with the metric $\rho(m^1, m^2) = sum_{ij}|(m^1_{ij} - m^2_{ij}|$. However, h^* is continuous if all cell boundaries have zero measure (Theorem 2.1 [6]).

If an ergodic μ lies in convex set $W \subset \mathcal{M}(f)$, then it must be an extreme point of W. A simple flow is generated by a cycle that cannot be decomposed into other flows. If such a flow lies in $V \subset \mathcal{M}(G)$, it must also be an extremal point of V.

Theorem 6.3 *Let $h^* : \mathcal{M}(f) \to \mathcal{M}(G)$ be a mapping from a set of invariant measures to flows on a graph. Then, the images $h^*(\mathcal{M}(f))$ and $h^*\left(\mathcal{M}^0\right)$ are convex subsets, and each extreme point in the image $h^*\left(\mathcal{M}^0\right)$ is the image of an ergodic measure*

$$ext\left(h^*\left(\mathcal{M}^0\right)\right) \subset h^*(ext(\mathcal{M}(f))).$$

Proof The mapping $h^* : \mathcal{M}(f) \to \mathcal{M}(G)$ is linear, so the images $h^*(\mathcal{M}(f))$ and $h^*\left(\mathcal{M}^0\right)$ are convex. The convex sets $\mathcal{M}(f)$ and $\mathcal{M}^0$ have the same extreme points, which are all ergodic. Each invariant measure $\mu \in \mathcal{M}^0$ can be decomposed into a finite sum of ergodic measures:

$$\mu = \sum_z \alpha_z \mu_z.$$

Consequently, the image of the measure μ is also decomposed into these images of ergodic measures:

$$h^*\mu = \sum_z \alpha_z h^* \mu_z.$$

Let $h^*\mu = m$ be an extreme point in $h^*\left(\mathcal{M}^0\right)$. The flow can be written as a sum of mages of ergodic measures $h^*\mu_z$

$$m = \sum_z \alpha_z h^* \mu_z.$$

For the extreme point m, this expansion is possible only if the image $h^*\mu_z$ for $\alpha_z > 0$ coincides with m, and the sum of such $\alpha_z > 0$ is 1. It follows that every extreme point in $h^*\left(\mathcal{M}^0\right)$ is the image of some extreme point in $\mathcal{M}^0$

$$ext\left(h^*\left(\mathcal{M}^0\right)\right) \subset h^*\left(ext\left(\mathcal{M}^0\right)\right) = h^*(ext(\mathcal{M}(f))),$$

so, the theorem is proved.

Preposition 6.1 *Let $h^* : \mathcal{M}(f) \to \mathcal{M}(G)$ be a mapping from the set of invariant measures to the set of flow on the graph G and the flow $m \in \mathcal{M}(G)$. Then*

(1) *the preimage $h^{*-1}(m) = \mathcal{M}(m)$ is a convex set;*
(2) *$\mathcal{M}^0(m) = \mathcal{M}^0 \cap \mathcal{M}(m)$ is a convex set;*
(3) *if m is an extreme point in the convex polyhedral $\mathcal{M}(G)$ and $\mathcal{M}(m) \neq \emptyset$, then the extreme points of $\mathcal{M}^0(m)$ are ergodic measures; that is,*

$$ext\left(\mathcal{M}^0(m)\right) \subset ext(\mathcal{M}(f)).$$

Proof

1. The mapping $h^* : \mathcal{M}(f) \to \mathcal{M}(G)$ is linear. If μ_1 and μ_2 lie in $\mathcal{M}(f)$, $h^*(\mu_1) = m = h^*(\mu_2)$, $\alpha + \beta = 1$ then $h^*(\alpha\mu_1 + \beta\mu_2) = m$; that is, $\mathcal{M}(m)$ is a convex set.
2. The set $\mathcal{M}^0(m)$ is convex as the intersection of convex sets $\mathcal{M}^0$ and $\mathcal{M}(m)$. If μ lies in $\mathcal{M}^0(m)$, then μ can be written as a finite sum of ergodic measures $\mu = \sum_k \alpha_k\mu_k$, and $h^*(\mu) = m$.
3. Let measure μ lie in $\mathcal{M}^0(m)$. Then μ can be written as a finite sum ergodic measures

$$\mu = \sum_k \alpha_k \, \mu_k, \quad \mu_k \in ext\mathcal{M}^0 = ext \; \mathcal{M}(f).$$

Let m be the extreme point of polyhedron $\mathcal{M}(G)$ and $h^*(\mu) = m$. We claim that all μ_k lie в $\mathcal{M}^0(m)$. In fact,

$$h^*(\mu) = \sum_k \alpha_k h^* \mu_k \quad \text{or} \quad m = \sum_k \alpha_k m_k,$$

where $h^*\mu_k = m_k$. Since m is an extreme point, the equality $m = \sum_k \alpha_k m_k$ is possible if $m_k = m$ and $\sum_k \alpha_k = 1$ for all k such that $\alpha_k > 0$. This means $h^*\mu_k = m$ and all ergodic measures μ_k lie in $\mathcal{M}^0(m) \subset \mathcal{M}(m)$. Let μ be an extreme point of $\mathcal{M}^0(m)$. By contradiction, show that μ is ergodic measure. If μ is not ergodic, then it could be written sum ergodic measures

$$\mu = \sum_k \alpha_k \, \mu_k, \quad \mu_k \in ext \; \mathcal{M}(f).$$

According to above, all μ_k lie in $\mathcal{M}^0(m)$ and therefore is not extreme in $\mathcal{M}^0(m)$. This contradiction completes proof.

Corollary *If $m \in h^*(\mathcal{M}(f))$ is simple flow, then m is image of some ergodic measure μ:*

$$m = h^*(\mu).$$

Preposition 6.2 *Let $\mathcal{M}(Q)$ be the set of flows on a directed graph Q, let G be another directed graph, and let $s : Q \to G$ be a mapping between directed graphs. Then the image $s^*(\mathcal{M}(Q))$ is a convex polyhedron in $\mathcal{M}(G)$, the vertices of which lie in the images of the vertices of the polyhedron $s^*(\mathcal{M}(G))$. That is,*

$$ext\big(s^*(\mathcal{M}(Q))\big) \subset s^*(ext(\mathcal{M}(Q))).$$

Proof According to Proposition 5.2, the mapping $s^* : \mathcal{M}(Q) \to \mathcal{M}(G)$ is linear and the image $s^*(\mathcal{M}(Q))$ is a convex polyhedron. Simple flows are the extreme points in the set $\mathcal{M}(Q)$ and the number of simple flows is finite. According to Theorem 5.1, every flow $m \in \mathcal{M}(Q)$ can be written as the sum of simple ones,

$$m = \sum_z \alpha_z m_z,$$

where $ext\,(\mathcal{M}(Q)) = \{m_z\}$ is the complete set of simple flows in $\mathcal{M}(Q)$. Then the image of the flow m has the form:

$$m^* = s^* m = \sum_z \alpha_z s^* m_z. \tag{6.2}$$

That is, every flow in the image $s^*(\mathcal{M}(Q))$ can be decomposed as a sum $\sum_z \alpha_z s^* m_z$. Let m^* be an extreme point in the convex set $s^*(\mathcal{M}(Q))$; that is, $m^* \in ext(s^*(\mathcal{M}(Q)))$. Then the flow m^* is decomposed into a sum of flows,

$$m^* = \sum_z \alpha_z s^* m_z.$$

Since m^* is an extreme point, this is only possible if all images $s^* m_z$ for $\alpha_z > 0$ are equal m^* (that is $s^* m_z = m^*$) and the sum of all such $\alpha_z > 0$ is 1. Consequently, every extreme point in the image $s^*(\mathcal{M}(Q))$ is the image of an extreme point of the set $\mathcal{M}(Q)$ so we have the inclusion:

$$ext\big(s^*(\mathcal{M}(Q))\big) \subset s^*(ext(\mathcal{M}(Q))).$$

This proves the proposition.

Let $s : Q \to G$ be a mapping of directed graphs. Consider a flow $m^* \in \mathcal{M}(G)$. There are two cases to consider:

(1) m^* is not the image of some flow under the mapping s^*; that is,

$$m^* \notin s^*(\mathcal{M}(Q)).$$

(2) m^* is the image of a flow under the mapping s^*; that is,

$$m^* \in s^*(\mathcal{M}(Q)).$$

Preposition 6.3 *Let $s : Q \to G$ be a mapping of directed graph, $m^* \in s^*(\mathcal{M}(Q))$. Then*

(1) *the preimage $s^{*-1}(m^*) = \mathcal{M}(m^*)$ is a convex subset of $\mathcal{M}(Q)$;*

(2) *if m^* is a simple flow on G and $\mathcal{M}(m^*) \neq \emptyset$, then the extreme points of the set $\mathcal{M}(m^*)$ are simple flows; that is,*

$$ext\big(\mathcal{M}(m^*)\big) \subset ext(\mathcal{M}(G)).$$

Proof

1. According to Proposition 5.2 the mapping $s^* : \mathcal{M}(Q) \to \mathcal{M}(G)$ is linear. If m_1 and m_2 lie in $\mathcal{M}(Q)$, $s^*(m_1) = m^* = s^*(m_2)$, $\alpha + \beta = 1$, then $s^*(\alpha m_1 + \beta m_2) = m^*$, and therefore $\mathcal{M}(m^*)$ is a convex set.
2. Let m be an extreme point of the set $\mathcal{M}(m^*) \subset \mathcal{M}(Q)$. The point m can be written as the sum of simple flows from the set $\mathcal{M}(Q)$:

$$m = \sum_k \alpha_k\ m_k, \quad m_k \in ext\ \mathcal{M}(Q).$$

Hence,

$$s^* m = \sum_k \alpha_k s^* m_k \quad or \quad m^* = \sum_k \alpha_k m_k^*,$$

where $s^* m_k = m_k^*$. By assumption, m^* is an extreme point of $\mathcal{M}(G)$. Consequently, $m^* = m_k^*$ for all k such that $\alpha_k > 0$. Therefore, all m_k lie in the preimage $s^{*-1}(m^*) = \mathcal{M}(m^*)$. The point m is an extreme point in $\mathcal{M}(m^*)$ and $m_k \in \mathcal{M}(m^*)$, and, therefore, it follows from the equality $m = \sum \alpha_k m_k$ that there exists k. such that $m_k = m$ and $\alpha_k = 1$. This means that m is an extreme point of the polyhedron $\mathcal{M}(Q)$. The proof is complete.

Corollary *If $m^* \in s^*(\mathcal{M}(Q))$ is a simple flow, then m^* is the image of a simple flow m:*

$$m^* = s^* m.$$

Consider successive subdivisions $C_1, C_2, C_3, \cdots$ such that the maximum cell diameter tends to zero. Suppose that each covering C_k consists of polyhedral cells intersecting at boundary disks. Let C be one of these coverings, and let $C^* = \{M^*(i)\}$ be the partition obtained from C by assigning each boundary disk of neighboring cells to one of those cells. In this case, a corresponding mapping $h(x) = \{i : x \in M^*(i)\}$ is uniquely defined. Without loss of generality, we assume that C_k^* are successive subdivisions of the original partition C_1^*. Each invariant measure μ generates a flow on the symbolic image defined by

$$h^* \mu = m = \big\{m_{ij} = \mu\big(M^*(i) \cap f^{-1}(M^*(j))\big)\big\}.$$

So, the flow $m^k = h_k^* \mu$ is defined on each symbolic image G_k. In this case, the constructed flows m^k are consistent:

$$s^* m^{k+1} = m^k.$$

Let m_k be an arbitrary flow on G_k. We have the following alternative: $m_k \notin s^* \mathcal{M}(G_{k+1})$ or $m_k \in s^* \mathcal{M}(G_{k+1})$.

In the first case, we say that the flow m_k has no extension over G_{k+1}; in the second, the flow m_k has an extension over G_{k+1}. According to the previous Corollary, if simple flow m_k has an extension, then there is a simple flow m_{k+1} such that

$$s^* m_{k+1} = m_k.$$

In other words, if a simple flow m_k has an extension, then this extension may be chosen to be simple. Suppose that on each symbolic image G_k a flow m^k is chosen and these flows are consistent, that is,

$$s^* m^{k+1} = m^k.$$

For each k, consider the measure μ_k on M defined by

$$\mu_k(A) = \sum_i m_i^k \frac{\nu(A \cap M(i))}{\nu(M(i))},$$

where ν is the Lebesgue measure.

Theorem 6.4 *Let C_k be successive subdivisions such that the maximal cell diameter d_k tends to zero. If m^k is a consistent sequence of simple flows on symbolic images G_k, then there exists an ergodic invariant measure*:

$$\mu = \lim_{k \to \infty} \mu_k.$$

First, we will prove the following statement.

Preposition 6.4 *Let $\mu_1 \neq \mu_2$ be invariant measures and let C_k be consecutive subdivisions such that the maximal diameter $d_k \to 0$. Then there exists a number t^* such that*

$$m_1^t = h_t^* \mu_1 \neq m_2^t = h_t^* \mu_2,$$

for all $t > t^*$, where $h_t^* : \mathcal{M}(f) \to \mathcal{M}(G_t)$ is the mapping from invariant measures to the space of flows on G_t.

Proof We will act by contradiction. Let $m_1^t = h_t^* \mu_1$, $m_2^t = h_t^* \mu_2$ and suppose that there exists a sequence $t \to \infty$, such that $m_1^t = m_2^t$. By construction, both the sequences m_1^t and m_2^t are consistent. For each t, let us define measures μ_1^t and μ_2^t on M setting

$$\mu_{12}^t(A) = \sum_i \left(m_{12}^t\right)_i \frac{v(A \cap M(i))}{v(M(i))},$$

where A is a measurable set, $M(i)$ are the cells in the division C_t, and v is the Lebesgue measure. As a result, we get two sequences of measures μ_1^t and μ_2^t on the manifold M. According to Theorem 5.3 invariant measures

$$\mu_1 = \lim_{t\to\infty} \mu_1^t, \quad \mu_2 = \lim_{t\to\infty} \mu_2^t$$

exist. Since, $\mu_1^t = \mu_2^t$ for each t and therefore $\mu_1 = \mu_2$. This contradiction proves the proposition.

Proof of Theorem 6.4 According to Theorem 5.3, there exists the weak limit

$$\mu = \lim_{k\to\infty} \mu_k$$

and the measure μ is invariant. Let us prove by contradiction that the measure μ is ergodic. Let μ be written as

$$\mu = \alpha\mu_1 + (1-\alpha)\mu_2,$$

where $\mu_1 \neq \mu_2$ and $0 < \alpha < 1$. Consider the mapping $h^* : \mathcal{M}(f) \to \mathcal{M}(G_k)$. We have

$$h^*\mu = \alpha h^*\mu_1 + (1-\alpha)h^*\mu_2 = \alpha m_1^t + (1-\alpha)m_2^t,$$

where $m_1^t = h^*\mu_1$, $m_2^t = h^*\mu_2$. Since the sequence m^k is consistent, we have

$$h^*\mu = m^k.$$

By Proposition 6.4, for all sufficiently large k, the flow $m_1^k \neq m_2^k$. Hence,

$$m^k = \alpha m_1^k + (1-\alpha)m_2^k,$$

where $0 < \alpha < 1$ and $m_1^k \neq m_2^k$. Therefore, the flow m^k is not simple for large k. This contradiction completes the proof.

6.3 Support of Invariant Measure

The support of the Borel measure μ is the set

$$supp(\mu) = \{x \in M : \mu(U) > 0 \text{ for any neighborhood } U, x \in U\}.$$

The following statement is proven in [2].

Preposition 6.5

(1) *The support of the Borel measure $supp(\mu)$ is closed.*
(2) *The complementary measure $\mu(M \backslash supp(\mu)) = 0$.*
(3) *Any set of full measure is dense in $supp(\mu)$.*

A set A has full measure if $\mu(A) = 1$. Consider the flow $m = \{m_{ij}\}$ on the symbolic image G. The measure of the vertex i is determined by the equality:

$$m_i = \sum_j m_{ij} = \sum_k m_{ki}.$$

The support of the flow m is the set of vertices:

$$supp(m) = \{i : m_i > 0\}.$$

By Theorem 5.1 each flow m is decomposed into the sum of simple flows:

$$m = \sum_k \alpha_k m(\omega_k),$$

where $\alpha_k \geq 0$, $\sum_k \alpha_k = 1$. Therefore, the support of the flow lies in the union of the supports of simple flows:

$$supp(m) = \bigcup_k suppm(\omega_k).$$

Preposition 6.6 *Let C be a covering of the manifold M by closed polyhedral cells intersecting along boundary disks. Suppose there exists a flow m on the symbolic image G and the measure μ is constructed according to a formula:*

$$\mu(A) = \sum_i m_i \frac{\nu(A \cap M(i))}{\nu(M(i))}.$$

Then the support of this measure is defined as

$$supp(\mu) = \left\{ \bigcup_i M(i) : i \in supp(m) \right\}.$$

Proof Consider a set of cells of the form:

$$Q = \left\{ \bigcup_i M(i) : i \in supp(m) \right\}.$$

Let $x \in Q$, then $x \in M(i)$ and $m_i > 0$. The Lebesgue measures of the cells are $\nu(M(i)) > 0$ by construction. If V is a neighborhood of the point x, then the Lebesgue measure $\nu(V \cap M(i)) > 0$ and

$$\mu(V) = \sum_i m_i \frac{\nu(V \cap M(i))}{\nu(M(i))} > 0.$$

Therefore, $x \in \; supp\ \mu$ and $Q \subset supp\,(\mu)$.

Now let us show the reverse inclusion. Let $x \in supp\mu$, then the measure $\mu(V) > 0$ for any neighborhood V of points x or

$$\sum_i m_i \frac{\nu(V \cap M(i))}{\nu(M(i))} > 0.$$

Therefore, there exists a cell $M(i)$ for which $m_i > 0$ and the Lebesgue measure $\nu(V \cap M(i)) > 0$. This means that the neighborhood V and the cell $M(i)$ intersect. Since V is an arbitrarily small neighborhood of the points x and $M(i)$ is a compact set, then the point x must be in $M(i)$. Therefore $x \in Q$ and $supp(\mu) \subset Q$. The proof is complete.

Finally, let C_k be sequential subdivisions with the maximal cell diameters converging to zero. If m^k is any consistent sequence of flows on the symbolic images G_k then according to Theorem 5.3 there is an invariant measure μ such that

$$\mu = \lim_{k \to \infty} \mu_k,$$

where convergence is considered in the weak topology.

Theorem 6.5 *Let C_k be successive subdivisions such that the maximal cell diameter $d_k \to 0$ and $\{m^k\}$ is a consistent sequence of flows on the symbolic images G_k, then the sequence of sets $Q_k = \{\bigcup_i M(i) : i \in supp(m^k)\}$ satisfies the following properties:*

1. *The sequence of compact sets Q_k monotonously decreases:*

$$Q_{k+1} \subset Q_k.$$

2. *The limit of the sequence of Q_k is the support of the invariant measure μ:*

$$\lim_{k\to\infty} Q_k = \bigcap_k Q_k = supp\ (\mu).$$

Proof

1. The equality $s(j) = i$ means that the cell $M^*_{k+1}(j)$ is included in the partition of $M^*_k(i)$. If $j \in supp(m^{k+1})$ and $s(j) = i$, then $M^*_k(i) \supset M^*_{k+1}(j)$ and

$$m^k_i = \mu(M^*_k(i)) \geq \mu(M^*_{k+1}(j)) = m^{k+1}_j > 0.$$

Therefore, $s(supp(m^{k+1})) \subset supp(m^k)$ and

$$Q_{k+1} = \left\{\bigcup M^*_{k+1}(j),\ j \in supp(m^{k+1})\right\} \subset \left\{\bigcup M^*_k(i),\ i \in\ supp(m^k)\right\} = Q_k.$$

2. Let be the point x in the intersection $\bigcap_k Q_k$ and U be a neighborhood of this point. Since the covering diameter converges to zero there exists k such that $x \in M^*_k(i) \subset U, i \in supp(m^k)$. Then

$$\mu(U) \geq \mu(M^*_k(i)) = m^k_i > 0.$$

Thus, $s(supp(m^{k+1})) \subset supp(m^k)$.

Let us prove the reverse inclusion by contradiction. Suppose that there exists a point $x \in supp(\mu)$, which does not belong to $\bigcap_k Q_k$. The set $\bigcap_k Q_k$ is compact set, so the distance $\rho(x, \bigcap_k Q_k) = r > 0$. Let V be a ball of radius $r/2$ with center at point x. We show that the measure $\mu(V) = 0$. Indeed, if $r > 0$, then there exists k for which the distance $\rho(Q_k, \overline{V}) > 0$. Then $m^k_i = 0$ if $M_k(i) \cap \overline{V} \neq \emptyset$. According to Proposition 5.2, the measure of a closed ball:

$$\mu(\overline{V}) = \lim_{k\to\infty} \sum_{i\in I} \mu(M^*_k(i)) = \lim_{k\to\infty} \sum_{i\in I} m^k_i,$$

where $I = \{i : M^*_k(i) \cap \overline{V} \neq \emptyset\}$. If k is large enough, then the values $m^k_i = 0$ and, hence, the measure $\mu(V)$ is also zero. The resulting contradiction proves the inclusion $supp(\mu) \subset \bigcap_k Q_k$. Thus, the theorem is proven.

This statement is proved in [2].

Preposition 6.7

1. *The support of an invariant measure lies in the set of recurrent points (trajectories).*
2. *If the measure μ is ergodic then there exists a dense trajectory in the support $supp\ \mu$.*
3. *If the restriction $f|_{supp\ \mu}$ is strictly ergodic, then $supp\ \mu$ is a minimal set.*

Recall that $f|_{supp\ \mu}$ is strictly ergodic when there is a single invariant measure on $supp\mu$. Recall that a closed invariant set is called minimal, if it does not have any invariant closed subsets. Generally speaking, minimal set can be support of more than one ergodic measure, see paragraph 10 of [7], an example of A.A. Markov on page 533 of [1], and the article [8].

Let C_k be sequential subdivisions with diameters d_k tending to zero, and let G_k be symbolic images for C_k. Suppose that simple flows m^k are chosen on G_k such that m^k are consistent. Theorem 6.4 guarantees the existence of an ergodic measure μ, which we can consider as the limit of the flows m^k. The simple flow m^k on G_k is generated by a simple cycle $\omega_k = \{i_1, i_2, \cdots, i_p = i_0\}$ with different vertices, $m_i = 1/p$ on ω_k, and $m_i = 0$ for all other vertices. Each periodic cycle ω_k generates a periodic pseudo-trajectory $T_k = \{x_i \in M_k(i) : i \in \omega_k\}$ according to Theorem 2.1.

Theorem 6.6 *Let C_k be sequential subdivisions. The cell diameter d_k converges to zero, and on each symbolic image G_k simple flows m^k are consistent. Then the following statements are true:*

1. *The periodic pseudo-trajectories T_k converge uniformly to a recurrent T trajectory.*
2. *The unions of cells of paths ω_k $Q_k = \{\bigcup_i M(i) : i \in \omega_k\}$ are nested:*

$$Q_k \supset Q_{k+1}.$$

3. *The sequence Q_k converges to the closure of the trajectory $\overline{T}$.*
4. *The limit $\lim_{k\to\infty} Q_k$ is a minimal set that is the support of the ergodic measure μ.*

Proof Theorem 3.1 (statements 1 and 2) guarantees that the pseudo-trajectories T_k will converge to the trajectory T. Moreover, the convergence is uniform because distances between T_k and T estimated through the diameter of the covering. Statement 4 of this theorem says that the trajectory T is recurrent and the limit $\lim_{k\to\infty} Q_k = \bigcap_k Q_k$ is a minimal set. The support for each simple flow m^k coincides with cycle ω_k. Then the set Q_k coincides with set

$$\left\{\bigcup M_k(i) : i \in supp\left(m^k\right)\right\}.$$

Theorem 6.4. states that a consistent sequence $\{m^k\}$ convergents to the ergodic measure μ. Theorem 6.5 states that the sequence of sets Q_k is monotonically decreasing, and its limit $\bigcup_k Q_k = \overline{T}$ is the support of measure μ. Theorem has been proven.

References

1. Nemytski, V.V., Stepanov, V.V.: Qualitative Theory of Dynamical Systems. Princeton University Press. Russian original 1949 (1960)
2. Katok, A., Hasselblatt, B.: Introduction to the Modern Theory of Dynamical Systems. Cambridge University Press (1995)
3. Sinai, YG.: Modern Problems of Ergodic Theory. Moscow (1995) (in Russian)
4. Billingsley, P.: Ergodic Theory and Information. Wiley, New York (1965)
5. Polovinkin, E.S., Balashova, M.V.: Elements of Convex and Strongly Convex Analysis. Moscow (2004) (in Russian)
6. Billingsley, P.: Convergence of Probability Measures. Wiley (1968)
7. Oxtoby, J.C.: Ergodic sets. Bull. Am. Math. Soc. **58**(2), 116–136 (1952)
8. Shtil'man, M.S.: Number of invariant measures with maximal entropy for translation in the space of sequences. Math. Notes **9**(3), 169–175 (1971)

Chapter 7
Entropy

7.1 Definitions

In 1865, R. Clausius introduced the concept of entropy into thermodynamics, a concept that has been a fundamental part of our understanding of the world for more than a century. L. Boltzmann, in 1872, pioneered a statistical approach to thermodynamics, which is based on the calculation of entropy. The Boltzmann entropy S of a system with n states is calculated as: $S = klog(n)$ where k is the Boltzmann constant. In 1948, C. Shannon introduced the concept of information channel capacity

$$C = \lim_{T \to \infty} \frac{log_2 N(T)}{T},$$

where $N(T)$ describes the maximum number of signals that can be transmitted through a channel in a certain amount of time T. He also defined information entropy as:

$$h = -\sum_i p_i log_2 p_i.$$

where $p(i)$ is the probability of the i-th signal (message), $i \in 1, ...n$ and n is the total number of signals.

A.N. Kolmogorov introduced entropy into the theory of dynamic systems in 1958. Entropy is an excellent invariant for a dynamic system and can be interpreted as a measure of chaos in the system.

For a comprehensive understanding of entropy in dynamic systems, see [1, 2]. It turns out that entropy characteristics can be obtained for both systems described analytically and for their phase portraits. The application of these characteristics to digital image analysis is described in [3].

G. Osipenko, *Symbolic Analysis of Dynamical Systems*, Mathematical Marvels: Texts and Monographs in the Spirit of CR Rao, https://doi.org/10.1007/978-981-95-0713-9_7

Motivation

Consider a discrete dynamical system $x_{n+1} = f(x_n)$ on a compact manifold M where $f : M \to M$ is a homeomorphism. Let $C = \{M(1), .., M(n)\}$ be a finite covering of M, and let $\left\{x_k = f^k(x), k = 0, \cdots, N-1\right\}$ be the segment of the trajectory. The covering C generates a coding of the trajectory through a finite sequence $\chi(x) = \{i_k, k = 0, \cdots N-1\}$, where $x_k \in M(i_k)$. In other words, i_k is the index of the cell containing the point. Generally speaking, the mapping $x_k \to i_k$ is multivalued. The sequence $\chi = \{i_k\}$ is called an admissible encoding of the trajectory $\left\{x_k = f^k(x)\right\}$ with respect to the covering C.

Let us assume that all admissible encodings with length N are known, and we need to predict subsequent encodings, that is, find admissible encodings with lengths $N+p$. Let $K(N)$ be the number of possible encodings that grow exponentially with N. We estimate the growth rate of codings using the formula:

$$h = \lim_{N\to\infty} \frac{log_b K(N)}{N}, \tag{7.1}$$

where b may be any real number larger than 1. The bases $b = 2$ (following Shannon) or $b = e$ are used. The existence of a limit in the above formula follows from following Lemma [4, 5].

Polya's Lemma 7.1 If a sequence of non-negative numbers satisfies the inequality

$$a_{n+m} \leq a_n + a_m,$$

Then there exists a limit for the sequence

$$\lim_{n\to\infty} \frac{a_n}{n}.$$

The number of allowable encodings satisfies the inequality

$$K(n+m) \leq K(n)K(m).$$

Consequently, there is a limit (7.1) for $a_n = log_b K(n)$. The exponential growth assumption (7.1) implies an estimation for the number $K(N)$

$$K(N) \sim Bb^{hN},$$

where B is a constant. If $h \neq 0$, then

$$\frac{K(N+p)}{K(N)} \sim b^{hp}.$$

This relation means that, for any N, the uncertainty of future encodings grows exponentially, regardless of knowledge of previous encodings of length N.

If the growth of different encodings is not exponential (that is $h = 0$), like $K(N) \sim BN^A$, then

$$\frac{K(N+p)}{K(N)} \sim \left(1 + \frac{p}{N}\right)^A \to 1,$$

when $N \to \infty$. In other words, the future ambiguity decreases as N increases.

Thus, if the increase in the number of encodings is exponential, uncertainty does not depend on N, otherwise it decreases with increasing N. The value of h can be interpreted as an indicator of uncertainty (chaos) in the system under study.

Topological Entropy.

Let f be a continuous mapping defined on a manifold M and $C = \{M(1), \cdots, M(n)\}$ be an open covering of M. For a positive integer N, consider a sequence

$$\omega = \omega_1 \omega_2 \cdots \omega_N,$$

where ω_k is a real number between 1 and n. Construct the intersection

$$M(\omega) = M(\omega_1) \cap f^{-1}(M(\omega_2)) \cap \cdots \cap f^{-N+1}(M(\omega_N)),$$

which is an open set. An admissible encoding ω corresponds to a nonempty intersection, that is, there exists $x \in M(\omega_1)$ such that $f^k(x) \in M(\omega_{k+1})$. The sequence ω codes a segment of the trajectory

$$\left\{f^k(x), k = 0, 1, \cdots, N-1\right\}.$$

Let's consider all the encodings $\{\omega\}$ of length N and a collection of sets $C^N = \{M(\omega)\}$, which forms an open covering. Let us choose a finite subcollection C_N in C^N that is minimal in terms of the number of elements (denoted below by $|C_N|$). Then, according to the Polya lemma, there exists a limit

$$H(C) = \lim_{N \to \infty} \frac{\log|C_N|}{N}. \tag{7.2}$$

Definition 7.1 *The number.*

$$h(f) = \sup_C H(C),$$

where supremum is over all open covers C, is called the topological entropy of the mapping $f : M \to M$.

It is easy to see that using this definition for the practical calculation of entropy makes little sense. Let's consider some methods for calculating entropy.

A covering C_2 is said to be refining in a covering C_1, if any element $A \in C_2$ lies in a set $B \in C_1$. A sequence of open coverings C_n is called exhaustive, if for any open covering C, there exists a number n^* such that the covering C_n is refined in C for $n \geq n^*$.

Proposition 7.1 [4]

1. *If C_n is a sequence of open coverings with diameters.*

$$d_n = \max_{A \in C_n} diam A$$

tending to zero, then C_n is an exhaustive sequence.

2. *The entropy of a mapping f is calculated as follows*

$$h(f) = \lim_{n \to \infty} H(C_n).$$

Consider coverings C_2 and C_1, and do not impose any special restrictions on the type of cells. For each covering, we construct nonempty intersections as described when defining the topological entropy:

$$M(\omega) = M(\omega_1) \cap f^{-1}(M(\omega_2)) \cap \cdots \cap f^{-N+1}(M(\omega_N)).$$

Here $M(i)$ are the cells of the considered covering. Let us denote the resulting collections of sets by C_1^N and C_2^N, respectively. In each collection, we select the minimum (by the number of elements) subcover and denote them C_{N1} and C_{N2}.

Proposition 7.2 *If C_2 is refined in C_1 then.*

$$|C_{N1}| \leq |C_{N2}|,$$

where $|C_{Ni}|$ denotes the number of elements in the set under consideration.

Proof If $A_1 \subset B_1$ and $A_2 \subset B_2$, then $A_1 \cap f^{-1}(A_2)$ lies in $B_1 \bigcap f^{-1}(B_2)$. In the same way, one can prove that the elements of C_2^N are in corresponding elements of C_1^N. Select the minimal subcovers C_{N1} and C_{N2}. Take all the elements from C_1^N which contain corresponding elements of C_{N2} and form a covering C_{N1}^*. Then $|C_{N1}| \leq \left|C_{N1}^*\right| \leq |C_{N2}|$, since C_{N1} is a minimal subcovering for C_1^N. The proposition is proven.

Corollary *Assume that C_2 is refined into C_1, the numbers $H(C_1)$ and $H(C_2)$ are calculated according to (7.2), then.*

$$H(C_1) \leq H(C_2)$$

7.2 Entropy of Symbolic Image

Let G be a graph with the adjacency matrix (the matrix of admissible transitions) Π. Let us denote the number of admissible n-length paths on G by b_n.

Definition 7.2 *The number*

$$h(G) = \lim_{n\to\infty} ln\frac{b_n}{n}$$

is called the entropy of the graph.

Remember that the element $(\Pi^n)_{ij}$ is equal to the number of admissible n-length paths from i to j. Then

$$b_n = \sum_{ij} (\Pi^n)_{ij}.$$

Theorem 7.1 *Let C_n be a sequence of subdivisions closed covering such cells with polyhedral intersecting boundary disks and let diameters d_n tend to zero. Let G_n denote the symbolic images of mapping f: $M \to M$ with respect to the sequence C_n. Then the entropy can be estimated as.*

$$h(f) \le \lim_{n\to\infty} h(G_n).$$

Proof Let $C_n = \{M(i)\}$ be a closed covering sequence described above, and G_n be a symbolic image of the mapping f constructed for C_n. Consider set P_N of encodings N-length segments trajectories. It is obvious that $|P_N|$ is not greater than number admissible N-length paths on G_n. Hence, the number $\left|C^N\right|$ of nonempty intersections.

$$M(\omega) = M(\omega_1) \cap f^{-1}(M(\omega_2)) \cap \cdots \cap f^{-N+1}(M(\omega_N)),$$

is not greater than b_N—the number N-length paths on G_n. Thus,

$$H(C_n) \le h(G_n).$$

If C is an open covering, then there exists the number n^* such that the covering C_n is refined in C for $n \ge n^*$. Then in accordance with Proposition 7.2

$$H(C) \le H(C_n) \le h(G_n).$$

Now consider an exhausting sequence of open coverings $\{\tilde{C}_m\}$. Let $n(m)$ be the number n^* constructed for the covering $\tilde{C}_m$, then

$$H(\tilde{C}_m) \le H(C_n) \le h(G_n),$$

where $n \geq n(m)$. If $m \to \infty$ then according to Proposition 7.1 we have

$$h(f) = \lim_{m \to \infty} H(\tilde{C}_m) \leq \lim_{n \to \infty} h(G_n).$$

Theorem is proven.

A matrix A with nonnegative (positive) elements is called nonnegative (positive). In particular, the matrix of admissible transitions Π is non-negative. A matrix $A(n \times n)$ is called decomposable if it admits an invariant subspace of dimension less than n. A matrix Π is indecomposable if the symbolic image consists of one class of equivalent recurrent vertices. If for a nonnegative matrix A there is an integer $s > 0$ such that all elements of the matrix A^s are positive, then the matrix A is called primitive.

Theorem 7.2 (Perron-Frobenius) [5, 6]

1. If A is an indecomposable matrix with nonnegative elements, then it has an eigenvector e with positive coordinates and a positive eigenvalue λ with multiplicity 1, and the other eigenvalues μ satisfy the inequality $|\mu| \leq \lambda$.
2. If A is an indecomposable matrix with nonnegative elements and $|\mu| < \lambda$, then A is primitive.
3. If A is an indecomposable matrix with nonnegative elements and has $h > 1$ eigenvalues which are equal in modulus λ, then A is not primitive and can be transformed into the form

$$\begin{pmatrix} 0 & A_{12} & 0 & \cdots & 0 & 0 \\ 0 & 0 & A_{23} & \cdots & \ddots & 0 \\ & & & & & \\ 0 & \ddots & 0 & \ddots & A_{h-1h} & \\ A_{h1} & 0 & 0 & \cdots & 0 & \end{pmatrix},$$

where A_{ij} are square blocks, and A^h consists of h primitive blocks.

Theorem 7.3 *The entropy of a graph G is equal to the logarithm of the maximum eigenvalue of the adjacency matrix.*

$$h(G) = ln\lambda.$$

Proof.

1. Consider a case where the graph G consists of one class of equivalent recurrent vertices. Let e be a positive eigenvector for the maximum eigenvalue λ,

$$\Pi e = \lambda e.$$

Then, $\Pi^n e = \lambda^n e$ and we have the following equality in the coordinate form

$$\sum_j (\Pi^n)_{ij} e_j = \lambda^n e_i. \tag{7.3}$$

Let $c = min\{e_i\}$ and $d = max\{e_i\}$. By the Perron-Frobenius theorem $c > 0$, and the following inequalities hold

$$c \sum_j (\Pi^n)_{ij} \leq \sum_j (\Pi^n)_{ij} e_j \leq d\lambda^n.$$

Which leads to the inequality

$$\sum_j (\Pi^n)_{ij} \leq \frac{d}{c}\lambda^n.$$

for any i. Summing over i we obtain:

$$b_n = \sum_{ij} (\Pi^n)_{ij} \leq \frac{dr}{c}\lambda^n,$$

where r is the number of rows of the matrix Π. From (7.3) it follows that

$$c\lambda^n \leq \lambda^n e_i = \sum_j (\Pi^n)_{ij} e_j \leq d \sum_j (\Pi^n)_{ij} \leq d \sum_{ij} (\Pi^n)_{ij}.$$

Hence, we get the estimation:

$$\frac{c}{d}\lambda^n \leq \sum_{ij} (\Pi^n)_{ij} = b_n.$$

Thus, we have the inequalities

$$\frac{c}{d}\lambda^n \leq b_n \leq \frac{dr}{c}\lambda^n.$$

Imply the required equality.

2 Consider the case where there are more than one class of equivalent recurrent vertices, that is, the matrix Π is decomposable. According to the Proposition 2. 2, by renumbering the vertices, this matrix can be transformed into the form:

$$\Pi = \begin{pmatrix} \Pi_1 & * & * & \cdots & * & * \\ 0 & \ddots & * & \cdots & * & * \\ 0 & 0 & \Pi_k & \ddots & \vdots & \vdots \\ \vdots & \ddots & \ddots & \ddots & * & \vdots \\ \vdots & \ddots & \ddots & \ddots & \ddots & * \\ 0 & 0 & \cdots & \cdots & 0 & \Pi_s \end{pmatrix},$$

where each diagonal block Π_k corresponds either to a class of equivalent recurrent vertices H_k, or to a nonrecurrent vertex. Under the diagonal blocks, there are zeros. Each class H_k has the entropy

$$h(H_k) = ln\lambda_k,$$

where λ_k is the maximum eigenvalue of the matrix Π_k. By the definition

$$h(G) = \lim_{n\to\infty} ln\frac{b_n}{n}.$$

Consider an admissible n-length path ω. Assume that on G there are s equivalence classes H_k. The path ω passes both through the vertices from H_k and nonrecurrent vertices that do not belong to these classes. Let us denote the parts of ω which lie in the classes H_k by ω_k. If we delete all the ω_k, it will contain only paths σ_l passing through nonrecurrent vertices. Thus, ω is the sum of ω_k and σ_l. Combine all the paths σ_l into a sequence σ. Generally speaking, it is not an admissible path. Let K be the number of nonrecurrent vertices in G. Then the sequence σ contains nonrecurrent vertices without repetitions. Therefore, the number of the sequences σ does not exceed the number of permutations of K elements and equals $K!$.

Denote the length of the path ω_k from the class H_k by $n(k)$. Then $n(1) + n(2) + \cdots + n(s) \leq n$. According to item 1 for each class H_k there is a number d such that the number $b_k(n(k))$ of different paths ω_k of length $n(k)$ is estimated as

$$b_k(n(k)) \leq d\lambda_k^{n(k)} \leq d\lambda^{n(k)},$$

where $\lambda = max\lambda_k$. Then, for the number of different paths ω_k belonging to ω, we have the estimate

$$\prod_k b_k(n(k)) \leq d^s\lambda^{n(1)+n(2)+\cdots+n(s)} \leq d^s\lambda^n.$$

Summing up the above estimates, we obtain the inequality

$$b_n \leq K!d^s\lambda^n.$$

Thus, we have an upper estimate for the entropy of a symbolic image:

$$h(G) \leq \lim_{n\to\infty} \frac{1}{n} ln\left[K!d^s\lambda^n\right] = ln\lambda.$$

Prove the opposite inequality. We note that the number of admissible paths on G is greater than the number of admissible paths in the class H_k. Then $h(G) \geq h(H_k) = ln\lambda_k$ for any k that gives a low estimate

$$h(G) \geq ln\lambda.$$

Therefore, we have

$$h(G) = ln\lambda$$

and the proof is finished.

7.3 Metric Entropy

Let μ be an invariant measure for the homeomorphism $f : M \to M$ and $C = \{A_1, A_2, \cdots, A_m\}$ a measurable partition of the manifold M.

Definition 7.3 *The entropy of the partition C is the number.*

$$H(C) = -\sum_i \mu(A_i)\ln \mu(A_i).$$

Let us construct a covering C^N which consists of the intersections of the form

$$A_{i_1} \cap f^{-1}\left(A_{i_2}\right) \cap \cdots \cap f^{-N+1}\left(A_{i_N}\right).$$

If such an intersection is not empty, then the sequence $i_1, i_2, \cdots i_N$ is an admissible coding with respect to the covering C. The metric entropy of f for the covering C is defined as

$$H(f, C) = \lim_{N\to\infty} \frac{1}{N} H\left(C^N\right).$$

The limit exists due to Polya's lemma.

Definition 7.4 *The entropy for the invariant measure μ is defined as*

$$h(f, \mu) = \sup_C H(f, C),$$

where supremum is over all measurable finite partitions.

Connection between topological and metric entropy is given by the following theorem.

Theorem 7.4 [7, 8] *Topological entropy of homeomorphism f is the upper bound of metrics entropies.*

$$h(f) = \sup_{\mu} h(f, \mu).$$

7.4 Stochastics Markov Chains

A stochastic Markov chain [5, 6] is defined by a collection of system states $\{i = 1, 2, \cdots n\}$ with probabilities P_{ij} of transition from state i to state j. The matrix of transition probabilities $P = \left(P_{ij}\right)$ is stochastic, that is, $P_{ij} \geq 0$ and $\sum\limits_{j} P_{ij} = 1$ for each i. A probabilistic distribution $p = (p_1, p_2, \cdots, p_n)$, $\sum\limits_{i} p_i = 1$ is stationary if it is left as a fixed vector for matrix:

$$(p_1, p_2, \cdots, p_n)\begin{pmatrix} P_{11} & P_{12} & \cdots & P_{1n} \\ P_{21} & P_{22} & \cdots & P_{2n} \\ \cdots & \cdots & \cdots & \cdots \\ P_{n1} & P_{n2} & \cdots & P_{nn} \end{pmatrix} = (p_1, p_2, \cdots, p_n),$$

It should be noted (see, [2]) that a stochastic matrix can be defined as the transpose of the described matrix, in which case a stationary distribution would be a right eigenvector.

There is a one-to-one correspondence between Markov chains and flows on the graph. Let $m = \left\{m_{ij}\right\}$ be a flow on the graph G. Measure of vertex i equals to $m_i = \sum_j m_{ij} = \sum_k m_{ki}$ If $m_i > 0$, then the vertex i must be recurrent. Each flow $m = \left\{m_{ij}\right\}$ on G generates a stochastic Markov chain with states $\{i\}$ being vertices $\{i : m_i \neq 0\}$ and transition probability $i \to j$ given by

$$P_{ij} = \frac{m_{ij}}{m_i}.$$

The resulting stochastic matrix $P = \left(\frac{m_{ij}}{m_i}\right)$ has the stationary distribution in the form of $(m_1, m_2, \cdots, m_n)$. This follows from the equality

$$(m_1, m_2, \cdots, m_n)\begin{pmatrix} \frac{m_{11}}{m_1} & \frac{m_{12}}{m_1} & \dots & \frac{m_{1n}}{m_1} \\ \frac{m_{21}}{m_2} & \frac{m_{22}}{m_2} & \dots & \frac{m_{2n}}{m_2} \\ \dots & \dots & \dots & \dots \\ \frac{m_{n1}}{m_n} & \frac{m_{n2}}{m_n} & \dots & \frac{m_{nn}}{m_n} \end{pmatrix} = (m_1, m_2, \cdots, m_n).$$

So, any flow $m = \{m_{ij}\}$ on a graph G generates a stochastic Markov chain for which the measure distribution (m_i) of vertices is stationary.

In turn, the converse is also true. For any stochastic matrix P and associated stationary distribution $p = (p_i)$, there exists a flow $m = \{m_{ij}\}$ on the graph G such that the measure distribution over the vertices is equal $m_i = p_i$. In fact, if P is a stochastic matrix $pP = p$, then there is a graph G with n vertices $\{i\}$ and edges $i \to j$, where $P_{ij} > 0$. Let us construct a measure on edges $m_{ij} = P_{ij}p_i$ and prove that it is a flow on G. Since P is a stochastic matrix, we have equality $\sum_j P_{ij} = 1$ for each i. Therefore

$$\sum_j m_{ij} = \sum_j P_{ij}p_i = p_i \sum_j P_{ij} = p_i.$$

Since $pP = p$, we have $\sum\limits_k p_k P_{ki} = p_i$, for each i, so we get

$$\sum_k m_{ki} = \sum_k p_k P_{ki} = p_i = \sum_j m_{ij},$$

that is, the Kirchoff's law holds for the distribution m_{ij}. Moreover $\sum_{ij} m_{ij} = \sum_i p_i = 1$. From the above, it follows that flow technology in a graph is equivalent to the stochastic-matrix method. In [9], a stochastic matrix of the form

$$P_{ij} = \frac{\nu\big(M(i) \cap f^{-1}(M(j))\big)}{\nu(M(i))},$$

is used, where ν is the Lebesgue measure and the $M(i)$ represent cells of a partition. In this article, the SBR-measure is constructed via the stochastic-matrix method under some additional conditions. It is clear that constructing a stochastic matrix with a non-Lebesgue measure leads, in general, to a different invariant measure then the SBR measure. For example, we can obtain a stochastic matrix from a zero–one matrix $\Pi = (\pi_{ij})$, where $\pi_{ij} = 1$ if the edge $i \to j$ exists, and $\pi_{ij} = 0$ otherwise. By setting,

$$P_{ij} = \frac{\pi_{ij}}{\sum_k \pi_{ik}},$$

we get a stochastic matrix, flow, and an approximation of the invariant measure.

7.5 Entropy of Flow

The technique developed can be applied to estimating metric entropy. Let $m = \{m_{ij}\}$ be a flow on a symbolic image G for the mapping f and the covering C. It was noted earlier that the flow m can be considered as an approximation of a certain invariant measure μ, if the cover diameter d is sufficiently small. A flow m on G generates a Markov chain with states being the vertices of G and transition probabilities are $p_{ij} = \frac{m_{ij}}{m_i}$. Under these conditions, the probability matrix $P = (p_{ij})$ has a stationary distribution $(m_1, m_2, \cdots, m_n)$. The entropy of the stationary distribution can be computed using the formula (see [5])

$$h_m = -\sum_i m_i \sum_j p_{ij} ln\, p_{ij}.$$

Substituting $p_{ij} = \frac{m_{ij}}{m_i}$, we find

$$h_m = -\sum_i m_i \sum_j \frac{m_{ij}}{m_i} ln \frac{m_{ij}}{m_i} = -\sum_{ij} m_{ij} ln \frac{m_{ij}}{m_i} =$$

$$-\sum_{ij} m_{ij} ln m_{ij} + \sum_{ij} m_{ij} ln m_i = -\sum_{ij} m_{ij} ln m_{ij} + \sum_i m_i ln m_i.$$

Thus, the entropy can be directly calculated via the flow m_i as

$$h_m = -\sum_{ij} m_{ij} ln m_{ij} + \sum_i m_i ln m_i. \tag{7.4}$$

The flow entropy with respect to the invariant measure μ can be estimated using the formula in (7.4), where the flow m approximates μ. Let's consider examples of calculating flow entropy.

Example 7.1 *Entropy of a simple flow.* Let $m = \{m_{ij}\}$ be a simple flow generated by a simple periodic path (cycle) $\omega = \{i_1, i_2, ..., i_p, i_{p+1} = i_1\}$, the vertices $i_1, i_2, ..., i_p$ are different. The flow m is such that $m_{ij} = {}^1/_p$ if the edge $i \to j$ belongs to ω; and $m_{ij} = 0$ otherwise. If a vertex $i \in \omega$, then there is only a single edge $j \to i \in \omega$ ending in i. Therefore, the measure $m_i = {}^1/_p$. If a vertex i does not lie on ω, then $m_i = 0$. The entropy of the simple flow.

$$h_m = -\sum_{ij} m_{ij} ln m_{ij} + \sum_i m_i ln m_i = lnp - lnp = 0.$$

Example 7.2 *Entropy of a flow with fixed branching.* Suppose that a graph G has n vertices and each vertex have k incoming edges and k outgoing edges. Therefore, the number of edges is equal to nk. There is a flow m on G such that.

$$m_{ij} = \frac{1}{nk}, m_i = \frac{1}{n}.$$

Therefore, the entropy of the flow m is

$$h_m = -\sum_{ij} m_{ij} ln m_{ij} + \sum_i m_i ln m_i = lnnk - lnn = lnk.$$

7.6 Flow of Maximal Entropy

Let Π be the transition matrix of a graph G. Our goal is to construct a flow with maximal entropy among all possible flows on G. Since each flow is grouped by recurrent vertices, we assume that G consists of a single component of recurrent vertices.

Theorem 7.5 *There exists a flow m on G such that.*

$$h_m = h(G) = ln\lambda.$$

Proof

1. The eigenvalues of the matrix A and its transpose (conjugate) matrix A^* coincide. Indeed, since $detA = detA^*$, then

$$det(A - \lambda E) = det(A - \lambda E)^* = det\left(A^* - \overline{\lambda}E\right).$$

Consequently, the eigenvalue λ of A corresponds to the conjugate eigenvalue $\overline{\lambda}$ of A^*. The roots of a real characteristic polynomial are either real or complex-conjugate, so the eigenvalues of the matrices A and A^* coincide.

2. Let Π be the admissible transition matrix of the graph G and λ be the maximal eigenvalue from the Perron-Frobenious theorem. Then for Π there exists a left eigenvector e with positive coordinates, $\sum_i e_i = 1$, such that

$$e\Pi = \lambda e, \Pi^* e = \lambda e.$$

Therefore, for each i we have

$$\sum_j \pi_{ji} e_j = \lambda e_i. \tag{7.5}$$

That leads to the equality

$$\sum_j \frac{\pi_{ji} e_j}{\lambda e_i} = 1$$

for every i. Therefore, a matrix of the form

$$P = \left(p_{ij} = \frac{\pi_{ji} e_j}{\lambda e_i} \right)$$

is a stochastic matrix with a left fixed vector m of stationary distribution:

$$mP = m.$$

According to the Sect. 7.4, the distribution on the arcs $i \to j$ defined by the equality

$$m_{ij} = p_{ij} m_i = \frac{\pi_{ji} e_j}{\lambda e_i} m_i, \tag{7.6}$$

is a flow m on the graph G such that the measure of the vertex i is equal to m_i. Entropy of the flow m is calculated by the formula

$$h_m = -\sum_{ij} m_{ij} \ln m_{ij} + \sum_i m_i lnm_i.$$

Hence,

$$h_m = -\sum_{ij} \frac{\pi_{ji} e_j}{\lambda e_i} m_i \ln \frac{\pi_{ji} e_j}{\lambda e_i} m_i + \sum_i m_i lnm_i.$$

Here we assume that $0ln0 = 0$. That means that the sum is over ij for which $\pi_{ij} = 1$. We have

$$-\sum_{ij} \frac{\pi_{ji} e_j}{\lambda e_i} m_i \ln \frac{\pi_{ji} e_j}{\lambda e_i} m_i = -\sum_{ij} \frac{\pi_{ji} e_j}{\lambda e_i} m_i \big(\ln \pi_{ji} + lne_j + \ln m_i - \ln \lambda - \ln e_i \big)$$

$$= \ln \lambda \sum_{ij} \frac{\pi_{ji} e_j}{\lambda e_i} m_i - \sum_{ij} \frac{\pi_{ji} e_j}{\lambda e_i} m_i \ln m_i + \sum_{ij} \frac{\pi_{ji} e_j}{\lambda e_i} m_i \big(\ln e_i - lne_j \big) =$$

$$\ln \lambda \sum_j \left(\sum_i \frac{\pi_{ji} e_j}{\lambda e_i} m_i \right) - \sum_i \left(\sum_j \frac{\pi_{ji} e_j}{\lambda e_i} \right) m_i \ln m_i + \sum_{ij} \frac{\pi_{ji} e_j}{\lambda e_i} m_i \big(\ln e_i - lne_j \big).$$

Taking into account the equalities

$$\sum_j \left(\sum_i \frac{\pi_{ji} e_j}{\lambda e_i} m_i \right) = \sum_j m_j = 1; \sum_j \frac{\pi_{ji} e_j}{\lambda e_i} = 1,$$

we get

$$h_m = \ln \lambda - \sum_i m_i \ln m_i + \sum_i m_i ln m_i + \sum_{ij} \frac{\pi_{ji} e_j}{\lambda e_i} m_i (\ln e_i - ln e_j)$$

$$= \ln \lambda + \sum_i \left(\sum_j \frac{\pi_{ji} e_j}{\lambda e_i} \right) m_i \ln e_i - \sum_j \left(\sum_i \frac{\pi_{ji} e_j}{\lambda e_i} m_i \right) ln e_j$$

$$= \ln \lambda + \sum_i m_i \ln e_i - \sum_j m_j ln e_j = \ln \lambda.$$

The proof is complete.

7.7 Calculation of Maximal Entropy Flow

Let's first consider at the method for calculating the left eigenvector for the maximum eigenvalue. This algorithm was proposed by Von Mises and Polachek-Goering in 1929 [10]. The algorithm is simple and exponentially convergent.

For simplicity, we will assume that the symbolic image has only one equivalence class of recurrent vertices. Since the algorithm reduces to sequential multiplication of a vector by a given matrix, it is well suited for large sparse matrices. Let Π be the matrix of admissible transitions for which we need to find the left eigenvector to the maximum eigenvalue. At the beginning of the algorithm, a random vector r_0 is generated. Next, iterative calculations are performed using the formula

$$r_{k+1} = \frac{r_k \Pi}{|r_k \Pi|}.$$

The distance between the elements in this sequence and the left eigenvector tends to zero. The sequence

$$\mu_k = \frac{(r_k, r_k \Pi)}{(r_k, r_k)},$$

where $(\cdot, \cdot)$ is the inner product, converges to the maximum eigenvalue. As a result of the algorithm, we get the maximum eigenvalue

$$\lambda_{max} = \lim_{k \to \infty} \mu_k$$

for the transition matrix Π and its left eigenvector

$$e = \lim_{k\to\infty} r_k = (e_1, e_2, ..., e_n).$$

Let's construct a stochastic matrix

$$P = \left(p_{ij} = \frac{\pi_{ji} e_j}{\lambda e_i} \right).$$

For the matrix P, we find the left fixed vector of the stationary distribution $m = (m_1, m_2, ..., m_n)$ such that

$$mP = m, m_k > 0, \sum_k m_k = 1.$$

This vector can be obtained by the algorithm described above. It is convenient to start with the initial vector equal to the vector e. For the constructed stochastic matrix P and its stationary distribution m, we define the flow

$$m_{ij} = p_{ij} m_i = \frac{\pi_{ji} e_j}{\lambda e_i} m_i.$$

According to the Theorem 7.5, the metric entropy reaches its maximum value on the constructed distribution

$$h_m = ln\lambda_{max}.$$

Example 7.3 Let's consider an example of how to calculate the maximum entropy flow. The flow was constructed for the Julia Mapping.

$$\begin{pmatrix} x \\ y \end{pmatrix} \to \begin{pmatrix} x^2 - y^2 + a \\ 2xy + b \end{pmatrix}.$$

We investigated the case where $a = 0, b = -\frac{1}{2\sqrt{2}}$.

The chain-recurrent set of the mapping has two components Ω_1 and Ω_2. The component Ω_1 is the boundary of the Julia set Q, which has a rather complex fractal structure. The other component Ω_2 is a stable fixed point inside Q. Trajectories in the region Q start at the unstable set Ω_1 and end at Ω_2. The system on Ω_1 has chaotic dynamics with nonzero entropy. We applied the technique above to calculate the invariant measure with maximum entropy. First, we localized the chain-recurrent set and constructed sufficiently small neighborhoods around components Ω_1 and Ω_2. Cells of the covering were squares with sides $\Delta = 0.00097656$. Figure 7.1 shows the neighborhoods of the components Ω_1 and Ω_2, which lie in the (xy) plane. These neighborhoods consist of 25 044 squares.

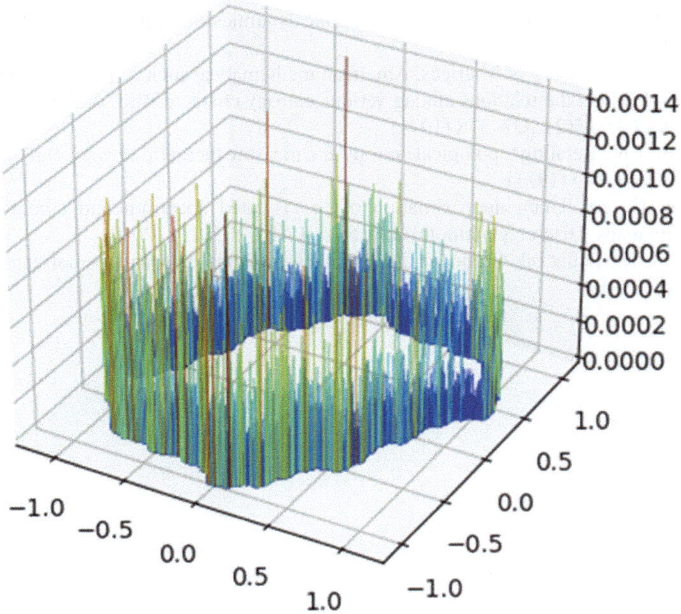

Fig. 7.1 The measure of the maximal entropy for the Julia mapping

Next, using the matrix Π of admissible transitions, we determined the left eigenvector e for the maximum eigenvalue of the matrix Π. Maximum entropy flow was calculated using the formula (7.6). Distribution of the measure is shown on the Fig. 7.1, the components Ω_1 and Ω_2 lie in the (xy)-plane, the values of m_i are plotted on the z-xis, where $m_i = \sum_j m_{ij}$ is the measure of the cell M (i). The metric entropy of the constructed flow was computed by formula (7.4) and amounted to E = 1.37241. To verify result obtained, the maximum eigenvalue $\lambda = 3.94487$ was calculated. Then the entropy of the symbolic image was equal to $\ln(\lambda) = 1.37242$, which differs from E by 0.00001. In conclusion, we have successfully calculated maximum entropy flow for the Julia Mapping.

References

1. Downarowicz, T.: Entropy in Dynamical Systems. Cambridge Univ, Press (2011)
2. Katok, A., Hasselblatt, B.: Introduction to the Modern Theory of Dynamical Systems. Cambridge University Press (1995)
3. Ampilova, N., Soloviev, I.: Entropies in the investigation of dynamical systems and their application to digital image analysis. J. Meas. Eng. **6**(2), 107–118 (2018). https://doi.org/10.21595/jme.2018.19891
4. Alekseev, V.M.: Symbolic Dynamics. 11th Mathematical School. Kiev (1976) (in Russian)

5. Lind, D., Marcus, B.: An introduction to symbolic dynamics and coding. Cambridge University Press (1995)
6. Gantmakher, F.: Theory of Matrices. American mathematical society, p. 374 (2000)
7. Dinaburg, E.I.: On the relations among various entropy characteristics of dynamical systems. Math. USSR-Izv. **5**(2), 337–378 (1971)
8. Goodman, T.N.: Relating topological entropy and measure theoretic entropy. Bull. Lond. Math. Soc. **3**(2), 176–180 (1971)
9. Froyland, G.: On Ulam approximation of the isolated spectrum and eigenfunctions of hyperbolic maps. Discret. Contin. Dynam. Syst. **17**(3), 203–221 (2007)
10. von Mises, R., Pollaczek-Geiringer, H.: Practical methods of equations solving. ZAMM **9**, 152–164 (1929)

Chapter 8
Averaging

8.1 Basic Definitions

Let us consider a discrete system

$$x_{n+1} = f(x_n). \tag{8.1}$$

generated by a homeomorphism $f : M \to M$ on a compact manifold M. Let φ be a continuous function, and let $\chi = \{x_k, 0 < k \leq n\}$ be a finite sequence of points. The number

$$\lambda(\chi) = \frac{1}{n}\sum_{k=1}^{n}\varphi(x_k).$$

is called the average value of φ over χ. If χ is a periodic sequence, then the average value $\lambda(\chi)$ is defined as the average over the period. For example, the average value of the function $\varphi(x) = \ln|det\, Df(x)|$ over a periodic trajectory, then its average value determines the rate of change of volume along that trajectory. Let $\chi = \{x_k, k = 1, 2, \ldots\}$ be an ε-semi-trajectory; if the limit

$$\lambda(\chi) = \lim_{n\to\infty}\frac{1}{n}\sum_{k=1}^{n}\varphi(x_k)$$

exists, it is called the averaging over the half-trajectory; if this limit does not exist, but there is some subsequence $n_m \to \infty$ for which the limit

G. Osipenko, *Symbolic Analysis of Dynamical Systems*, Mathematical Marvels: Texts and Monographs in the Spirit of CR Rao, https://doi.org/10.1007/978-981-95-0713-9_8

$$\lambda(\chi, n_m) = \lim_{n_m \to \infty} \frac{1}{n_m} \sum_{k=1}^{n_m} \varphi(x_k) \tag{8.2}$$

exists, then this subsequence is said to be convergent.

Definition 8.1 *The averaging spectrum of the function φ along the ε-semitrajectory $\chi = \{x_k, k = 1, 2, \ldots.\}$ is a set of averages of convergent subsequences.*

$$\Sigma(\chi) = \{\lambda : \text{there is } n_m \to \infty \text{ such that } \lambda = \lambda(\chi, n_m)\}.$$

If χ is a periodic ε-trajectory, then all these limits exist and they are equal to the average value over the period.

Definition 8.2 *The spectrum of the averaging of φ over pseudo-trajectories of the dynamical system (8.1) is defined as follows.*

$$\Sigma = \Big\{\lambda : \text{ there exists a sequence of } \varepsilon_n - \text{trajectories } \{\chi_n\},$$
$$\varepsilon_n \to 0, \quad \lambda_n \in \Sigma(\chi_n), \quad \lambda = \lim_{n\to\infty} \lambda_n\Big\}$$

Thus, the spectrum Σ is the limit set of average values over ε-trajectories as $\varepsilon \to 0$.

8.2 Properties of Spectrum

Proposition 8.1 *Let $\chi = \{x_k, k \in N^+\}$ be an ε-semitrajectory. Assume that there exists a convergent subsequence.*

$$\lambda(\chi, n_m) = \lim_{n_m \to \infty} \frac{1}{n_m} \sum_{k=1}^{n_m} \varphi(x_k).$$

Then there exists a sequence of periodic 2ε-trajectories $\{\chi_p\}$, $p \to \infty$, such that

$$\lambda(\chi_p) \to \lambda(\chi, n_m) \text{ as } p \to \infty.$$

Proof The limit $\lambda(\chi, n_m)$ does not depend on the finite number of initial values $\varphi(x_k)$, that is, instead of the sequence $\chi = \{x_k, k > 0\}$ we can consider the sequence $\chi(N_0) = \{x_k, k \geq N_0\}$. In this case,

$$\lambda(\chi, n_m) = \lim_{n_m \to \infty} \frac{1}{n_m - N_0} \sum_{k=N_0}^{n_m - 1} \varphi(x_k), \text{ where } n_m > N_0.$$

Since the sequence $\{x_{n_m}\}$ lies in a compact set, there exists a subsequence converging to some point x^*. Without loss of generality, we can assume that the original sequence $\{x_{n_m}\}$ itself converges to x^*. By the assumption, the original sequence χ is an ε-trajectory. Let U be the $\varepsilon/2$-neighborhood of the point x^*. There exists a number N_0 such that all points $\{x_{n_m}, n_m \geq N_0\}$ lie in U, and hence, the distance between them is less than ε. So, we can show that the finite sequence $\chi_p = \{x_k, N_0 \leq k < n_p\}$ is a periodic 2ε-trajectory. Indeed, by the assumption $\{x_k\}$ is an ε-trajectory, and therefore we need to estimate the distance between $f(x_{n_p-1})$ and x_{N_0}. By construction $\rho(x_{N_0}, x_{n_p}) < \varepsilon$ and $\rho(f(x_{n_p-1}), x_{n_p}) < \varepsilon$. Therefore, $\rho(f(x_{n_p-1}), x_{N_0}) < 2\varepsilon$. The averaging over $\chi_p = \{x_k, N_0 \leq k < n_p\}$ is

$$\lambda(\chi_p) = \frac{1}{n_p - N_0} \sum_{k=N_0}^{n_p-1} \varphi(x_k).$$

Thus, $\lambda(\chi, n_m) = \lim_{p\to\infty} \lambda(\chi_p)$ completing the proof.

The spectrum Σ can be obtained as a limit set of average over periodic ε-trajectories as $\varepsilon \to 0$; that is,

$$\Sigma = \{\lambda : \text{there exists a sequence of periodic } \varepsilon_n - \text{trajectories},$$
$$\varepsilon_n \to 0, \text{ such that } \lambda = \lim_{n\to\infty} \lambda(\chi_n)\Big\}.$$

Consider a sequence $\{\chi_n\}$ of periodic ε_n-trajectories, where $\varepsilon_n \to 0$, such that $\lambda = \lim_{n\to\infty} \lambda(\chi_n)$. Suppose that there is a sequence of points $x_n \in \chi_n$ converging x^*. According to Proposition 3.4, the point x^* is chain-recurrent. Let Ω be a component of the chain recurrent set containing point x^*. According to Theorem 3.6, sequence $\{x_n\}$ converges uniformly to the component Ω. In this case, we will say that the averaging λ is realized on the component Ω.

Proposition 8.2 *Let $\{\chi_n\}$ be a sequence of periodic pseudo-trajectories with the above properties and let x^* be its limit point. Then there exists sequence $\{\omega_n\}$ of periodic ε_n-trajectories passing through x^*, $\varepsilon_n \to 0$, such that $\lambda(\omega_n) \to \lambda$.*

Proof Consider sequence of periodic ε_n^*-trajectories $\{\omega_n\}$, constructed in the proof of Proposition 3.4. By the construction, each pseudo-trajectory ω_n is obtained from χ_n by replacing point x_n by limit point x^*. Besides this, ω_n is a periodic ε_n^*-trajectory, where $\varepsilon_n^* \to 0$. Distance $\rho(x_n, x^*) < \delta_n$ tends to zero as $n \to \infty$. If $\chi_n = \{x_1, x_2, \cdots, x_k = x_0\}$ is a periodic trajectory with mark point x_0, which coincides with x_n, then $\omega_n = \{x_1, x_2, \cdots, x_k^* = x_0^*\}$, where $x_0^* = x^*$ and $\rho(x_0, x_0^*) = \rho(x_n, x^*) = \delta_n$. Then averaging becomes

$$\lambda(\chi_n) = \frac{1}{k}\sum_{i=0}^{k-1}\varphi(x_i) = \frac{1}{k}\left(\sum_{i=1}^{k-1}\varphi(x_i) + \varphi(x^*)\right) + \frac{1}{k}(\varphi(x_n) - \varphi(x^*))$$
$$= \lambda(\omega_n) + E_n,$$

where $E_n = \frac{1}{k}(\varphi(x_n) - \varphi(x^*)) \to 0$ as $n \to \infty$. Proof is complete.

Proposition 8.3 *If λ_1 and λ_2 are realized on a component Ω, then any number λ between λ_1 and λ_2 is also realized on Ω.*

Proof By assumption, there are two sequences of periodic ε-trajectories $\{\chi_n\}$ and $\{\psi_n\}$ such that.

$$\lambda_1 = \lim_{n\to\infty} \lambda(\chi_n),$$

$$\lambda_2 = \lim_{n\to\infty} \lambda(\psi_n).$$

The limit sets of these sequences lie in Ω. Therefore, there are two sequences of points $\{x_n \in \chi_n\}$ and $\{y_n \in \psi_n\}$ which converge to the points $x^* \in \Omega$ and $y^* \in \Omega$, respectively. In view of Proposition 8.2, it can be assumed that the periodic pseudo-trajectories $\{\chi_n\}$ pass through x^* and the periodic pseudo-trajectories $\{\psi_n\}$ pass through y^*. Since x^* and y^* lie in the component Ω, for any $\varepsilon_n > 0$ there is a periodic ε_n-trajectory χ^* passing through x^* and y^*. The transition from y^* to x^* along χ^* will be called the first half of χ^*; the transition from y^* to x^* will be called the second half of χ^*. Let us construct a periodic pseudo-trajectory ξ_n from three periodic pseudo-trajectories χ^*, χ_n and ψ_n. Starting from the point x^*, make N rounds along χ_n, and then go through the first half of the trajectory χ^* to the point y^*. Then we make K rounds along ψ_n and return to the starting point x^* along the second half of the trajectory χ^*. Averaging over ξ_n depends on the number of circuits N and K, which can be arbitrarily large. The averaging over the linking pseudo-trajectory χ^* is fixed. Any number λ between λ_1 and λ_2 can be written as

$$\lambda = \alpha\lambda_1 + (1-\alpha)\lambda_2,$$

where $0 < \alpha < 1$. The period of the trajectory ξ_n is equal to $NT_1 + KT_2 + T_3$, where T_1, T_2 and T_3 are the periods of χ_n, ψ_n and χ^* respectively. The average over ξ_n is given by the expression

$$\lambda(\xi_n) = \frac{1}{NT_1 + KT_2 + T_3}\sum_{z=0}^{NT_1+KT_2+T_3-1}\varphi(x_z). \tag{8.3}$$

Which breaks down into 3 terms. The first term is generated by N circuits along pseudo trajectory χ_n, the second term is generated from K circuits along pseudo-trajectory ψ_n, and the last term is generated by the trajectory χ^*. By collecting like terms, we can rewrite the average as

$$\lambda(\xi_n) = \frac{1}{NT_1 + KT_2 + T_3}\big(NT_1\lambda(\chi_n) + KT_2\lambda(\psi_n) + T_3\lambda(\chi^*)\big).$$

Now we choose numbers N and K such that

$$\frac{NT_1}{NT_1 + KT_2 + T_3} \to \alpha, \frac{KT_2}{NT_1 + KT_2 + T_3} \to (1-\alpha), \quad \frac{T_3}{NT_1 + KT_2 + T_3} \to 0,$$

as $N \to \infty$ and $K \to \infty$. To do this, we construct sequences of integers V and W such that $V, W \to \infty$ and $V/W \to \alpha$. We have $0 < V < W$, because $0 < \alpha < 1$. Let define sequences of numbers N and K such that

$$\frac{NT_1}{NT_1 + KT_2} = \frac{V}{W}.$$

For example, we can put

$$N = T_2V, K = T_1(W - V). \tag{8.4}$$

Given fixed T_1, T_2 and T_3, we have

$$\frac{T_3}{NT_1 + KT_2 + T_3} \to 0,$$

and

$$\frac{NT_1}{NT_1 + KT_2 + T_3} = \frac{NT_1}{NT_1 + KT_2}\frac{NT_1 + KT_2}{NT_1 + KT_2 + T_3} \to \alpha$$

as $N \to \infty$ and $K \to \infty$. Let sequences $N_n \to \infty$ and $K_n \to \infty$ be chosen as indicated above. Thus, we have a sequence ξ_n such that

$$\lambda(\xi_n) = \alpha\lambda(\chi_n) + (1-\alpha)\lambda(\psi_n) + \delta_n,$$

where $\delta_n \to 0$. As a result,

$$\lambda(\xi_n) \to \alpha\lambda_1 + (1-\alpha)\lambda_2,$$

as $n \to \infty$. Proof is complete.

Theorem 8.1. *The averaging spectrum Σ of a function φ consists of intervals $[a_k, b_k]$, where each interval is generated by a component of the chain-recurrent set Ω_k, $a_k = \lambda_{inf}(\Omega_k)$ and $b_k = \lambda_{sup}(\Omega_k)$ are the lower and upper bounds of averaging on Ω_k.*

Proof From Propositions 3.4, 3.5, and 8.2 it follows that every $\lambda \in \Sigma$ is realized on some component Ω. Having fixed this component Ω, we consider all λ that are realized on Ω and set $a = inf\{\lambda\}$. We claim that a is realized on Ω. Indeed, there exists a sequence λ_k converging to a as $k \to \infty$ such that each λ_k is realized on Ω. For any k, there is a sequence ω_{nk} of periodic ε_{nk}-trajectories such that $\lambda(\omega_{nk}) \to \lambda_k$ and $\varepsilon_{nk} \to 0$ as $n \to \infty$. For any $\varepsilon > 0$, there exists k^* such that $|\lambda_{k^*} - a| < \varepsilon/2$. We fix k^* and for such k^* we find n^* such that $|\lambda_{k^*} - \lambda(\omega_{n^*k^*})| < \varepsilon/2$ and $\varepsilon_{n^*k^*} < \varepsilon$. Therefore, $|a - \lambda(\omega_{n^*k^*}))| < \varepsilon$. Since ε is arbitrary, a lies in the spectrum. According to Proposition 8.2, we can assume that the periodic pseudo-trajectories ω_{nk} pass through the point $x^* \in \Omega$. From Proposition 3.5 it follows that a is realized on Ω, and a similar analysis shows that $b = sup\{\lambda\}$ is realized on Ω. The Proposition 8.3 guarantees that the interval $[a, b]$ lies in Σ, which proves the assertion.

8.3 Averaging on Graph

Let G be a directed graph. A function on G is a distribution (framing) $\{a[ij]\}$ on the edges $\{i \to j\}$ of the graph. Averaging of the function over a finite path $\omega = \{z_0, z_1, ..., z_p\}$ is the arithmetic mean

$$\lambda(\omega) = \frac{1}{p}\sum_{k=1}^{p} a[z_{k-1}z_k].$$

In particular, if $\omega = \{z_0, z_1, ..., z_p = z_0\}$ is a periodic path, then $\lambda(\omega)$ is the average value of the function on the period. If $\omega = \{z_k, k = 0, 1, 2, \cdots\}$ is an infinite path, then the limit

$$\lambda(\omega) = \lim_{n\to\infty} \frac{1}{n}\sum_{k=1}^{n} a[z_{k-1}z_k]$$

is called the averaging over ω, if the limit exists. In the case when this limit does not exist, then there is a subsequence $n_m \to \infty$, for which the limit

$$\lambda(\omega, n_m) = \lim_{n_m\to\infty} \frac{1}{n}\sum_{k=0}^{n_m-1} a[z_k z_{k+1}] \tag{8.5}$$

exists, we call it converging.

Definition 8.3 *The averaging spectrum over* ω *is the limit set of averages over converging subsequences.*

$$\Sigma(\omega) = \{\lambda : \textit{there is}\, n_m \to \infty\, \textit{such that}\, \lambda = \lambda(\omega, n_m)\}.$$

The averaging spectrum over G is the limit set of the averages over all admissible paths

$$\Sigma(G) = \left\{\lambda : \lambda = \lim_{m\to\infty} \lambda_m, \quad \lambda_m \in \Sigma(\omega_m), \quad \omega_m \text{ are admissible paths}\right\}.$$

Thus, the spectrum Σ(G) is the closure of the union of spectra over all spectra.

$$\Sigma(G) = \overline{\bigcup_{\omega} \Sigma(\omega)}$$

Proposition 8.4 *Let* $\omega = \{z_k, k = 0, 1, 2, \cdots\}$ *be an infinite path on G. Suppose that there exists a convergent subsequence (8.5). Then there exists a sequence of periodic paths* $\{\omega_p\}, p \to \infty$ *such that* $\lambda(\omega_p) \to \lambda(\omega, n_m) npu\, p \to \infty$.

Proof The limit (8.5) does not depend on the first few terms of the sequence ω, that is,

$$\lambda(\omega, n_m) = \lim_{n_m\to\infty} \frac{1}{n - n_0} \sum_{k=n_0}^{n_m-1} a[z_k z_{k+1}].$$

Since the graph has a finite number of vertices, there is some vertex $z^* = z_{n_0} = z_{n_p}$, that appears infinitely often in ω. Then the finite sequence $\omega_p = \{z_k, n_0 \le k \le n_p\}$ is a periodic path on G. Averaging over ω_p gives

$$\lambda(\omega_p) = \frac{1}{n_p - n_0} \sum_{k=n_0}^{n_p-1} a[z_k z_{k+1}].$$

Thus, $\lambda(\omega, n_m) = \lim_{p\to\infty} \lambda(\omega_p)$, and this proves the statement.

Corollary. *The* $\Sigma(G)$ *spectrum coincides with the set of limits of averages over periodic paths:*

$$\Sigma(\mathrm{G}) = \left\{\lambda \text{ there exists a sequence of periodic paths } \omega_\mathrm{n} \text{ such that } \lambda = \lim_{\mathrm{n}\to\infty} \lambda(\omega_\mathrm{n})\right\}.$$

Thus, the spectrum of a graph is determined by its recurrent vertices.

Consider the sequence of periodic paths $\{\omega_n\}, n \to \infty$, on which the limit $\lambda \in \Sigma(G)$ is reached. The recurrent set RV is divided into a finite number $\{H_m\}$

of equivalence classes of recurrent vertices. If the sequence $\{\omega_k\}$ lies in the class H and $\lambda = \lim_{k\to\infty} \lambda(\omega_k)$, then we say that the value λ is realized in the class H. The spectrum of the class H is defined as

$$\Sigma(\mathrm{H}) \\ = \left\{\lambda : \text{ there exists a sequence of periodic paths } \omega_n \subset \mathrm{H} \text{ such that } \lambda = \lim_{n\to\infty} \lambda(\omega_n)\right\}.$$

Proposition 8.5 *If λ_1 and λ_2 are realized in the class H, then any number λ between λ_1 and λ_2 is also realized in that class H.*

The proof of this proposition repeats the proof of Proposition 8.3.

Proposition 8.6 *The spectrum of the class H is an interval $[a(H), b(H)]$, where $a(H) = \lambda_{inf}(H)$ and $b(H) = \lambda_{sup}(H)$ are the lower and upper bounds for averaging over periodic paths.*

Proof Set $b = sup\{\lambda \in \Sigma(H)\}$. Then there is a sequence $\lambda_k \in \Sigma(H)$ such that $\lambda_k \to b$ as $k \to \infty$. For each k there is a sequence of periodic paths ω_{nk} in H such that $\lambda(\omega_{nk}) \to \lambda_k$ as $n \to \infty$, and the diagonal sequence of paths ω_{nn} generates the averaging sequence $\lambda_n = \lambda(\omega_{nn})$, converging to b as $n \to \infty$. Consequently, b lies in the spectrum $\Sigma(H)$. Similarly, a lies in the spectrum $\Sigma(H)$. It follows from the previous propositions that the segment $[a, b]$ lies in $\Sigma(H)$. From the definition of the numbers a and b, it follows that $[a, b] = \Sigma(H)$. The proof is over.

Consider a class H of equivalent recurrent vertices and a periodic path $\omega \subset H$. A periodic path $\omega = \left\{z_1, ..., z_p = z_0\right\}$ is called a simple path or a cycle if its vertices $z_1, ..., z_p$ are different. The number of simple paths is finite, because the number of vertices is also finite. We claim that the path ω can be represented as the sum of simple paths. Indeed, if the path ω is not simple, then it has a repeated vertex z^*: $z^* = z_l = z_{l+p_1}, 0 < p_1 < p$. Consider two periodic sequences $\omega^* = \left\{z_1, ..., z_{l-1}, z_{l+p_1}, ..., z_p = z_0\right\}$ and $\omega^{**} = \left\{z_{l+1}, ..., z_{l+p_1} = z_l\right\}$. Since the arcs $z_{l-1} \to z_l = z_{l+p_1}$ and $z_{l+p_1} = z_l \to z_{l+1}$ exist, the sequences ω^* and ω^{**} are admissible periodic paths with periods p_1 and $p_2 = p - p_1$, respectively. It is clear that $p_1, p_2 < p$. In this case, we say that the path ω is the sum of periodic paths ω^* and ω^{**}:

$$\omega = \omega^* + \omega^{**}.$$

Repeatedly decomposing, we arrive at a decomposition of ω in the sum of periodic paths $\omega_1, ..., \omega_q$ with periods $p_1, ..., p_q, p_1 + ... + p_q = p$. Taking into account that periods are integers, we complete the decomposition process and obtain a decomposition $\omega = \varphi_1 + \varphi_2 + ... + \varphi_r$ consisting of simple periodic paths. Note that the simple paths $\varphi_1, \varphi_2, ..., \varphi_r$ can coincide. When a periodic path ω^* repeats k times in ω, we write

$$\omega = k\omega^* + \omega^{***}.$$

Symbolic image has finite vertex set and so the number of simple periodic paths is finite. Let $\varphi_1, \varphi_2, ..., \varphi_q$ be the family of all simple periodic paths in the class H with periods $p_1, ..., p_q$, respectively. Then periodic path $\omega \subset H$ can be written as

$$\omega = k_1\varphi_1 + k_2\varphi_2 + ... + k_q\varphi_q$$

Without loss, assume each simple periodic path φ_j is contained in ω with coefficient $k_j \geq 0$ (the case $k_j = 0$ means that ω does not pass through φ_j).

Let $\varphi_1, \varphi_2, ..., \varphi_q$ be the set of all cycles of the class H with periods $p_1, ..., p_q$. Find the averages over these cycles

$$\lambda(\varphi_j) = \frac{1}{p_j}\sum_{k=1}^{p_j} a\left[z_{k-1}^j z_k^j\right].$$

Let $\omega \subset H$ be some periodic path with its decomposition $\omega = k_1\varphi_1 + k_2\varphi_2 + ... + k_q\varphi_q$ and its period $p = k_1p_1 + k_2p_2 + ... + k_qp_q$. In this case, we say that a simple periodic path φ_j enters ω with weight $\mu_j = \frac{k_jp_j}{p}$. It is clear that $\sum_{j=1}^{q} \mu_j = 1$.

Proposition 8.7 *The averaging over a periodic path $\omega = k_1\varphi_1 + k_2\varphi_2 + ... + k_q\varphi_q$ can be calculated as*

$$\lambda(\omega) = \sum_{j=1}^{q} \mu_j\lambda(\varphi_j).$$

Proof Let $\omega = \{i_1, ..., i_p\}$ be a periodic path in the symbolic image G. By definition, the averaging over the periodic path ω is

$$\lambda(\omega) = \frac{1}{p}\sum_{r=1}^{p} a[i_r i_{r+1}].$$

Grouping the terms along the periodic paths φ_s, we get

$$\lambda(\omega) = \frac{1}{p}\sum_{s=1}^{q} k_s \sum_{l=1}^{p_s} a[i_l i_{l+1}] = \frac{1}{p}\sum_{s=1}^{q} k_s p_s \lambda(\varphi_s) = \sum_{s=1}^{q} \frac{k_s p_s}{p}\lambda(\varphi_s) = \sum_{s=1}^{q} \mu_s\lambda(\varphi_s).$$

Thus, the averaging over $\lambda(\omega)$ is a weighted average of simple periodic paths φ_s with weights μ_s. The proof is complete.

Let

$$\lambda_{min}(H) = min\{\lambda(\varphi_j), j = 1, ..., q\},$$

$$\lambda_{max}(H) = max\{\lambda(\varphi_j), j = 1, ..., q\}$$

be the minimum and maximum of the averages over simple periodic paths in the class H. From the Proposition 8.7 it follows

Proposition 8.8 *The averaging spectrum $\Sigma(H)$ of the class H is the interval*

$$\Sigma(H) = [\lambda_{min}(H), \lambda_{max}(H)]$$

Theorem 8.2 *The spectrum $\Sigma(G)$ of the averaging over graph G consists of the intervals $[\lambda_{min}(H), \lambda_{max}(H)]$, where $\{H_k\}$ is the complete family of classes of equivalent recurrent vertices on the graph G.*

Proof According to Proposition 8.8, the spectrum $\Sigma(G)$ contains the intervals $[\lambda_{min}(H_k), \lambda_{max}(H_k)]$, where $\{H_k\}$ is the complete set of classes of equivalent recurrent vertices. We need to prove that there are no other values in the spectrum. Suppose that there exists $\lambda \in \Sigma(G)$ that does not belong to the intervals $[\lambda_{min}(H), \lambda_{max}(H)]$. By the definition of the spectrum, there exists a sequence of periodic paths ω_n such that.

$$\lambda = \lim_{n\to\infty} \lambda(\omega_n). \tag{8.6}$$

Each periodic path ω_n belongs to some class H. If the class H contains a finite number of paths from the sequence ω_n, then the limit (8.6) is independent of these paths. The number of classes H is finite. Therefore, a class H exists that contains an infinite number of paths in the sequence ω_n. These paths form a subsequence ω_{n_k}, $n_k \to \infty$. Any subsequence of a convergent sequence has the same limit. Therefore, the subsequence $\lambda(\omega_{n_k})$ converges to λ. Since all ω_{n_k} are in the class H, then $\lambda \in [\lambda_{min}(H), \lambda_{max}(H)]$, which contradicts the assumption about the position of the number λ. The theorem is proven.

8.4 Estimation of Averaging Spectrum

Consider a symbolic image G of a dynamical system $x_{n+1} = f(x_n)$ and a continuous function $\varphi : M \to R$. To calculate the averaged spectrum of the function over pseudo-trajectories, we construct a function on a symbolic image by assigning a value $a[ij]$ to each edge $i \to j$ as follows. Let $i \to j$ be an edge in G. Fixing a point x in a cell $M(i)$, we assign $a[ij] = \varphi(x)$. Note that the value $a[ij]$ does not depend on j. If x^* is another point in $M(i)$ and $a^*[ij] = \varphi(x^*)$, then we have the estimate

$$|a^*[ij] - a[ij]| = |\varphi(x) - \varphi(x^*)| < \theta(d),$$

where $\theta(d)$ is the modulus of continuity of the function φ, d is the diameter of the covering. Let us consider the constructed function on the symbolic image. According to the previous theorem, we can find the averaged spectrum of the symbolic image in a finite number of steps. The task is to estimate averaged spectrum Σ of function φ through the spectrum of the symbolic image $\Sigma(G)$.

Theorem 8.3

1. *Averaged spectrum Σ of function φ lies in the extended spectrum Σ^* of the symbolic image. The extended spectrum consists of the intervals.*

$$[\lambda_{min}(H_k) - \theta(d), \lambda_{max}(H_k) + \theta(d)],$$

where $\{H_k\}$ is the complete family of equivalence classes of recurrent vertices, d is the diameter of the covering, and $\theta(d)$ is the modulus of continuity of function φ.
2. *If the diameter d converges to 0, then the extended spectrum Σ^* and spectrum of the symbolic images $\Sigma(G)$ converge to Σ in Hausdorff metrics.*

Hausdorff distance is defined as follows: let X and Y be the subsets of a compact space K and let $\rho(x, y)$ be a distance on K. The distance between X and Y is defined as

$$dist(X, Y) = \sup\{\rho(x, Y), \rho(y, X) : x \in X, y \in Y\}$$

where $\rho(x, Y) = inf\{\rho(x, y) : y \in Y\}$. If $X \subset Y$, then $\rho(x, Y) = 0$ and $dist(X, Y) = sup\{\rho(y, X) : y \in Y\}$.

Proof of statement 1. Let λ lie in the spectrum averaging of the function φ. Claim λ lies in extended spectrum of the symbolic image. Indeed, λ lies in the spectrum therefore there is a sequence of periodic ε_n-trajectories $\{\chi_n\}$, $\varepsilon_n \to 0$ such that $\lambda = \lim \lambda(\chi_n)$ as $\varepsilon_n \to 0$. Let $\chi_n = \{x_1, x_2, \cdots, x_p = x_0\}$. By Theorem 2.1, if $\varepsilon_n < r$, then the periodic path $\omega_n = \{i_k : x_k \in M(i_k)\}$ is admissible in the symbolic image. Let

$$\lambda(\chi_n) = \frac{1}{p}\sum_{k=1}^{p} \varphi(x_k)$$

and

$$\lambda(\omega_n) = \frac{1}{p}\sum_{k=1}^{p} \varphi(x_k^*),$$

where $|\varphi(x_k) - \varphi(x_k^*)| < \theta(d)$. We have

$$\lambda(\chi_n) = \lambda(\omega_n) + \delta_n$$

where $\delta_n < \theta(d)$. Note that these inequalities hold uniformly for all sufficiently large n. Moreover, for large n periodic paths ω_n lie in some class H of equivalent recurrent vertices. This means that $\lambda(\chi_n) \in [\lambda_{min}(H) - \theta(d), \lambda_{max}(H) + \theta(d)]$ and $\lambda \in [\lambda_{min}(H) - \theta(d), \lambda_{max}(H) + \theta(d)]$. The proof is complete.

Proof of assertion 2. The spectrum of a symbolic image $\Sigma(G)$ lies in its extended spectrum Σ^*. The Hausdorff distance between these two spectra is $\theta(d)$ and $\theta(d) \to 0$ as $d \to 0$. We have $\Sigma \subset \Sigma^*$, and

$$\Sigma \subset \lim_{d\to 0} \Sigma^* = \lim_{d\to 0} \Sigma(G).$$

Therefore $dist(\Sigma^*, \Sigma) = sup\{\rho(\lambda, \Sigma) : \lambda \in \Sigma^*\}$, let us show by contradiction that $\lim_{d\to 0} dist(\Sigma^*, \Sigma)$ is zero. Assume that it is not true. Then there would exist sequences d_n and λ_n such that $d_n \to 0$, $\lambda_n \in E^*$, but $\rho(\lambda_n, \Sigma) > q > 0$; that is, λ_n is uniformly separated from the spectrum. The inclusion $\lambda_n \in E^*(G_n)$ means that on the symbolic image G_n there is a periodic path ω_n, $\lambda_n = \lambda(\omega_n) + \varepsilon_n$, $|\varepsilon_n| \le \theta(d_n)$. By Theorem 2.1, this periodic path ω_n generates a periodic ε-trajectory σ_n for any $\varepsilon > d_n$ (for example, for $\varepsilon = \frac{3}{2}d_n$). Moreover, $\lambda(\sigma_n) = \lambda(\omega_n) + \delta_n$, $|\delta_n| < \theta(d_n)$. Without loss of generality, we can assume that all the considered sequences converge. Hence $\lambda = lim\lambda(\sigma_n) = lim\lambda(\omega_n)$ and $\lambda(\omega_n)$ is bounded away from the spectrum. Therefore, λ does not lie in Σ, which contradicts the definition of the spectrum. Thus, the theorem is proven.

To estimate the average spectrum of a system, we must find the spectrum of the framed symbolic image. This problem reduces to finding cycles with minimum and maximum averages. It is clear that finding such cycles using brute-force approach is impossible in our case, as the number of cycles increases rapidly during iterative construction of a symbolic image. Therefore, we can use the method of the special transportation problem in linear programming [1] or the algorithm for finding the minimum mean cycle [2].

8.5 Averaging and Flows on Graph

Let us consider an abstract directed graph G on which the function $a[ij]$ is defined. Let $m = \{m_{ij}\}$ be a flow on G. We call the quantity

$$b(m) = \sum_{ij} a[ij]m_{ij}$$

the averaging over the flow m. According to the Proposition 5.4, each periodic path $\omega = \{z_0, z_1, \cdots, z_p = z_0\}$ generates a flow of the form

$$m(\omega) = \left\{ m_{ij} = \frac{k_{ij}(\omega)}{p} \right\},$$

where $k_{ij}(\omega)$ is the number of times the path passes ω through the arc $i \to j$. Such a flow is called periodic. Let us now compare the averaging $\lambda(\omega)$ over a periodic path ω and the averaging $b(m)$ over flow $m(\omega)$ generated by periodic path ω.

Proposition 8.9 *Let H be the equivalence class of recurrent vertices. The set of periodic averages.*

$$\Delta_{per}(H) = \left\{ \lambda(\omega) = \frac{1}{p} \sum_{k=1}^{p} a\big[z_k z_{k+1}\big], \omega = \big\{z_0, z_1, \cdots, z_p = z_0\big\} \subset H \right\}$$

coincides with set of averages over periodic flows

$$\Delta_{flow}(H) = \left\{ b(m) = \sum_{ij} a\big[ij\big] m_{ij}, \big\{m_{ij}\big\} \text{ is a periodic flow on H} \right\}.$$

Proof Consider periodic path $\omega = \big\{z_0, z_1, \cdots, z_p = z_0\big\}$ on H. Grouping similar terms in the arithmetic mean $\lambda(\omega)$, we obtain.

$$\lambda(\omega) = \frac{1}{p} \sum_{k=1}^{p} a\big[z_k z_{k+1}\big] = \frac{1}{p} \sum_{ij} k_{ij}(\omega) a\big[ij\big] = \sum_{ij} \frac{k_{ij}(\omega)}{p} a\big[ij\big],$$

where $k_{ij}(\omega)$ is the number of passes of the path ω through the arc $i \to j$. According to Proposition 5.4, the distribution $m(\omega) = \left\{ m_{ij} = \frac{k_{ij}(\omega)}{p} \right\}$ is flow generated by ω. We have the equality

$$\lambda(\omega) = b(m).$$

Thus, for any periodic path ω, the mean value on it is the same as averaging over the flow $m(\omega)$. The proof is complete.

Theorem 5.2 states that periodic flows are dense in space of flows.

Corollary. *The averaging spectrum $\Sigma(H)$ of the class H is the closure of averages over periodic flows on H:*

$$\Sigma(H) = \overline{\Delta_{flow}(H)}.$$

Theorem 8.4 *Let G be a directed graph and $\big\{a\big[ij\big]\big\}$ be a function defined on G. Then its averaged spectrum $\Sigma(G)$ is the union of all intervals*

$$\Sigma(G) = \bigcup_{k} [a_k, b_k],$$

where each $[a_k, b_k] = \Sigma(H_k)$ *is a set of average values over flows* $\mathcal{M}(H_k)$, *and* $\{H_k\}$ *is a complete set of equivalent recurrent vertex classes.*

Proof It follows from the Theorem 8.2 that we need only prove this statement for each equivalence class of recurrent vertices. Consider a class of recurrent vertices H. The equality $\lambda(\omega) = b(m(\omega))$ holds for every periodic path $\omega \subset H$, where $m(\omega)$ is the flow generated by ω. Let $\lambda \in \Sigma(H)$, so there is a sequence of periodic paths ω_n such that $\lambda(\omega_n)$ converges to λ. We need to show that there is some flow m^* such that $\lambda = b(m^*)$. The sequence of flows $m(\omega_n)$ lies in a compact set, so there is a subsequence ω_{n_k} that $m(\omega_{n_k})$ converges to m^* in the metric ρ. Then the sequence $b(m(\omega_{n_k}))$ converges to $b(m^*)$ as $k \to \infty$. So, we have.

$$\lambda = \lim_{k\to\infty} \lambda(\omega_{n_k}) = \lim_{k\to\infty} \lambda(\omega_{n_k}) = \lim_{k\to\infty} b(m(\omega_{n_k})) = b(m^*).$$

Thus, theorem is proven.

8.6 Averaging and Invariant Measures

Let Ω be a component of a chain-recurrent set and $\mathcal{M}(\Omega)$ be the set of normalised invariant measures concentrated on Ω. Consider a continuous function $\varphi : M \to R$ and a measure $\mu \in \mathcal{M}(\Omega)$. We say that

$$\int_M \varphi d\mu = \lambda(\mu)$$

is the averaging of a function φ with respect to a measure μ. The following theorem establishes a connection between the averaging spectra and invariant measures.

Theorem 8.5 *The spectrum of the averaging over pseudo-trajectories of the dynamic system is the union of intervals.*

$$\Sigma = \bigcup_k [a_k, b_k],$$

where each interval $[a_k, b_k] = \Sigma(\Omega_k)$ *is a set of averages over invariant measures concentrated in the component* Ω_k *of the chain-recurrent set:*

$$\Sigma(\Omega_k) = \left\{ \lambda = \int_M \varphi d\mu, \ \mu \in \mathcal{M}(\Omega_k) \right\},$$

where $\{\Omega_k\}$ *is the complete family of chain-recurrent components.*

Proof To prove the theorem, it is enough to show that.

$$\Sigma(\Omega) = \left\{ \lambda = \int\limits_M \varphi d\mu, \mu \in \mathcal{M}(\Omega) \right\},$$

for each component of the chain-recurrent set Ω, where $\Sigma(\Omega) = [a, b]$ is the averaging spectrum interval generated by the component Ω.

Part I. Let Ω be a component of the chain-recurrent set. Considering an invariant measure $\mu \in \mathcal{M}(\Omega)$, we show that $\int_M \varphi d\mu$ lies in the interval Σ. For a fixed $\varepsilon > 0$, we claim that there is a periodic pseudo-trajectory $\omega = \{x_i, i = 1, 2, \cdots, p\}$, for which the difference between the arithmetic mean $b(\omega) = \frac{1}{p}\sum_{i=1}^{p} \varphi(x_i)$ and the averaging with respect to the measure $\int_M \varphi d\mu$ is less than ε. The pseudo-trajectory ω will be constructed as follows. On the symbolic image G the measure μ generates a flow m (by formula (5.1). By Theorem 5.2 near the flow m there is a periodic flow m^* generated by a periodic path ω^* on G, and the path ω^* generates the desired periodic pseudo-trajectory ω according to the weak shadowing property (Theorem 2.1).

Let $C = \{M(i)\}$ be a covering of the manifold M whose cells are closed polyhedra intersecting in boundary discs. Let G be the symbolic image with respect to the covering C. Fix a point x_i inside each cell $M(i)$, and $b[i] = \varphi(x_i)$. There is a unique equivalence class of recurrent vertices $H(\Omega)$ for the component Ω such that $U = \{\cup M(i), i \in H(\Omega)\}$ is a closed neighborhood of the set Ω. If the diameter of the covering d tends to zero, then $U \to \Omega$. Using the covering $C = \{M(i)\}$, we can construct a partition $C^* = \{M^*(i)\}$, composed of polyhedra whose boundary discs lie in the same neighboring cell. We can also construct a flow m

$$m_{ij} = \mu\Big(f\big(M^*(i)\big) \bigcap M^*(j)\Big) = \mu(M^*(i) \bigcap f^{-1}\big(M^*(j)\big)$$

on the symbolic image. The measure μ is concentrated on $\Omega \subset \left\{\bigcup M(i), i \in H(\Omega)\right\}$, which means that the flow m is concentrated on $H(\Omega)$. The quantity

$$\mu\big(M^*(i)\big) = \sum_j m_{ij} = \sum_k m_{ki} = m_i$$

defines the measure of the vertex i. The averaging of the function $b[i]$ over the flow m_{ij} is the number

$$\sum_{ij} m_{ij} b[i] = \sum_i m_i b[i] = \sum_i \mu\big(M^*(i)\big)\varphi(x_i).$$

Thus, we have obtained an integral sum, which depends on the choice of the points x_i in $M(i)$. By the mean value theorem, we have

$$\int\limits_{M} \varphi(x)d\mu = \sum_{i} \int\limits_{M^*(i)} \varphi(x)d\mu = \sum_{i} \varphi(x_i^*)\mu(M^*(i)),$$

where the internal point x_i^* lies in the cell $M(i)$. So,

$$\left|\sum_{i} m_i b[i] - \int\limits_{M} \varphi(x)d\mu\right| = \left|\sum_{i} \varphi(x_i)\mu(M^*(i)) - \sum_{i} \varphi(x_i^*)\mu(M^*(i))\right| \leq \theta(d). \tag{8.7}$$

where $\theta(d)$ is the modulus of continuity of φ and d is the diameter of the covering. By choosing d small enough, the difference between the averaging of a flow and the averaging with respect to measurement can be made sufficiently small. Periodic flows are dense in the space of flows $\mathcal{M}(\Omega)$. This means that for any $\delta > 0$, there is a periodic flow $m^* = \left\{m_{ij}^*\right\}$ such that $\sum_{ij} |m_{ij} - m_{ij}^*| < \delta$. Flow m^* is generated by a periodic path $\omega^* = \{i_k, k = 1, 2, \cdots, p\}$ so that $m_{ij}^* = \frac{k_{ij}}{p}$, where k_{ij} is the number of time path ω^* passes edge $i \to j$. We have

$$\left|\sum_{i} m_i b[i] - \sum_{i} m_i^* b[i]\right| = \left|\sum_{i} b[i] \sum_{j} (m_{ij} - m_{ij}^*)\right| \leq K \sum_{ij} \left|m_{ij} - m_{ij}^*\right| < K\delta,$$

where $K = max|\varphi|$. Averaging over periodic flow m^* has the form

$$\sum_{i} m_i^* b[i] = \sum_{ij} m_{ij}^* b[i] = \frac{1}{p} \sum_{ij} k_{ij} b[i] = \frac{1}{p} \sum_{k} \varphi(x_{i(k)}),$$

where $x_{i(k)} \in M(i_k), k = 1, 2, \cdots, p$. Averaging over flow m^* coincides with averaging over the periodic path:

$$\frac{1}{p} \sum_{k} \varphi(x_{i(k)}) = b(\omega).$$

According to Theorem 2.1 (as applied to a closed covering), there is a periodic sequence of points $\omega = \{x_k, k = 1, 2, \cdots, p\}$ for the periodic path ω^* such that $x_k \in M(i_k)$ and ω is an ε_1-trajectory for any $\varepsilon_1 > d$. We assume $\varepsilon_1 = \frac{3}{2}d$. Since $x_{i(k)}$ and x_k both lie in $M(i_k)$, have

$$\left|\frac{1}{p} \sum_{k} \varphi(x_{i(k)}) - \frac{1}{p} \sum_{k} \varphi(x_{i_k})\right| < \theta(d).$$

Summing the above inequalities, we have

$$\left|\int_M \varphi d\mu - b(\omega)\right| = \left|\int_M \varphi d\mu - \frac{1}{p}\sum_k \varphi(x_k)\right| \le \left|\int_M \varphi d\mu - \sum_i m_i b[i]\right|$$
$$+ \left|\sum_{ij} m_{ij} b[i] - \sum_{ij} m^*_{ij} b[i]\right|$$
$$+ \left|\sum_{ij} m^*_{ij} b[i] - \frac{1}{p}\sum_k \varphi(x_k)\right| < 2\theta(d) + K\delta.$$

Let $d > 0$ and $\delta > 0$ be such that $2\theta(d) + K\delta \le \varepsilon$. In this case, the difference between the averaging $b(\omega)$ over the periodic pseudo-trajectory ω and averaging with measure $\int_M \varphi d\mu$ is less than ε.

Approach described above allows us to construct a sequence of periodic ε_n-trajectories $\{\omega_n, \varepsilon_n \to 0\}$, the average values of which $b(\omega_n)$ converge to the averaging with respect to the measure $\int_M \varphi d\mu$. Moreover, ω_n lies in the neighborhood $U(\Omega)$, and ω_n converges to Ω as $d \to 0$. Set of averages over the measures $\left\{\int_M \varphi d\mu, \mu \in \mathcal{M}(\Omega)\right\}$ lies in the interval $\Sigma(\Omega)$.

Part II. Let us prove converse inclusion. Need show that for any element $\lambda \in \Sigma(\Omega)$ of spectrum there exists invariant measure μ with support in Ω such that $\lambda = \int_M \varphi d\mu$. Consider a sequence of periodic ε_n-trajectories $\omega_n = \{x_k, k = 1, 2, \cdots, p_n\}$, $\varepsilon_n \to 0$ such that the averages $b(\omega_n) = \frac{1}{p_n}\sum_k \varphi(x_k)$ converge to λ. By assumption, λ is realized on a component Ω, it means that the sequence $\{\omega_n\}$ converges uniformly to Ω by Theorem 3.6. Let $C = \{M(i)\}$ be a covering by cells, which are polyhedra intersecting in boundary discs, and let G be the symbolic image of f with respect to C. Let $H(\Omega)$ be unique equivalence class of recurrent vertices such that $U = \{\cup M(i), i \in H(\Omega)\}$ is a closed neighborhood of the set Ω. According to Theorem 2. 1, if ε_n is sufficiently small, sequence $\sigma = \{i_k, x_k \in M(i_k)\}$ is an admissible path in the graph G. Moreover, if n is large enough, then ω_n lies in the neighborhood $U = \{\cup M(i), i \in (H(\Omega)\}$ and the periodic path σ lies in H. Let $\{b[i] = \varphi(x_i), x_i \in M(i)\}$ define function on G and let $b(\sigma) = \frac{1}{p_n}\sum_k \varphi(x_k)$ be the arithmetic mean over σ. Then

$$|b(\omega_n) - b(\sigma)| < \theta(d).$$

The periodic path σ generates the flow

$$m = \left\{m_{ij} = \frac{k_{ij}}{p_n}\right\},$$

where k_{ij} is the number of times the path σ traverses the edge $i \to j$. In this case, the averaging over the flow $b(m) = \sum_i m_i b[i]$ coincides with $b(\sigma)$. The flow m generates a measure μ such that

$$\mu(A) = \sum_i m_i \frac{\nu(A \cap M(i))}{\nu(M(i))},$$

where $m_i = \sum_j m_{ij}$ is the measure of a cell $M(i)$ and ν is the Lebesgue measure. By the construction, the measure μ is concentrated on $U = \{\cup M(i), i \in H(\Omega)\}$. Averaging with respect to this measure gives

$$\int_M \varphi d\mu = \sum_i \int_{M^*(i)} \varphi d\mu = \sum_i \varphi(x_i^*) m_i,$$

where x_i^* is an interior point in $M(i)$. Therefore,

$$\left| \int_M \varphi d\mu - b(m) \right| = \left| \sum_i \varphi(x_i^*) m_i - \sum_i \varphi(x_i) m_i \right| < \theta(d)$$

and

$$\left| \int_M \varphi d\mu - b(\omega_n) \right| \le \left| \int_M \varphi d\mu - b(m) \right| + |b(\sigma) - b(\omega_n)| < 2\theta(d).$$

Consider a sequence of coverings C_k with diameters $d_k \to 0$. Using the technique described above, construct a sequence of measures μ_k. According to Theorem 5.6, there exists a subsequence $\mu_{k(t)}$ converging to an invariant measure μ_0 (in the weak topology). By Theorem 3.9 (on localizing) the neighborhoods U converge to the component Ω as $d \to 0$. Therefore, the measure μ_0 concentrats on Ω. The original sequence of pseudo-trajectories $\{\omega_n\}$ contains a subsequence $\{\omega_{n(t)}\}$ corresponding to $\{\mu_{n(t)}\}$. Without loss of generality, we assume that $\{\omega_n\}$ corresponds to the sequence $\mu_n \to \mu_0$. So, we have sequence $\{\omega_n\}$ of periodic ε_n-trajectories such that

$$\left| \int_M \varphi d\mu_n - b(\omega_n) \right| \to 0 \text{ as n} \to \infty,$$

where $\varepsilon_n \to 0$, $b(\varepsilon_n) \to \lambda$, the sequence of measures μ_n converges weakly to μ_0, and μ_0 is an invariant measure. Therefore, $\int\limits_M \varphi d\mu_n \to \int\limits_M \varphi d\mu_0$ as $n \to \infty$, and $\int\limits_M \varphi d\mu_0 = \lambda$. The theorem is proven.

8.7 Extremal Property of Ergodic Measures

The set $\mathcal{M}(\Omega)$ of invariant measures concentrated on a component of the chain-recurrent set Ω is a convex weakly compact set. The ergodic measures form the set of extreme points $ext\mathcal{M}(\Omega)$ of $\mathcal{M}(\Omega)$. By Choquet's Theorem, any point in a convex closed set is an integral sum of its extreme points. Recall that the set of measures which are a convex combination of a finite number of ergodic measures

$$\mathcal{M}^0 = \left\{ \mu = \sum_{k=1}^{m} \alpha_k \mu_k \mu_k \in ext\, \mathcal{M}(\Omega), \alpha_k \geq 0, \sum_k \alpha_k = 1 \right\},$$

is convex, and its closure in the weak topology coincides with $\mathcal{M}(\Omega)$,

$$\overline{\mathcal{M}^0} = \mathcal{M}(\Omega).$$

Let us fix a continuous function φ and consider the functional

$$\Phi(\mu) = \int\limits_M \varphi d\mu$$

on the space of measures $\Phi : \mathcal{M}(\Omega) \to R$. This functional is linear in μ and continuous. If $\mu = \alpha\mu_1 + (1 - \alpha)\mu_2, \alpha > 0$, then the definition of the Lebesgue integral implies the equality

$$\int\limits_M \varphi d\mu = \alpha \int\limits_M \varphi d\mu_1 + (1 - \alpha) \int\limits_M \varphi d\mu_2.$$

We have

$$\Phi(\mu) = \alpha\Phi(\mu_1) + (1 - \alpha)\Phi(\mu_2). \tag{8.8}$$

The functional Φ is continuous in the weak topology and it maps the convex compact set $\mathcal{M}(\Omega)$ into a convex compact set in R, namely the closed interval $[a, b]$,

$$a = \min_{\mathcal{M}(\Omega)} \Phi(\mu), \quad b = \max_{\mathcal{M}(\Omega)} \Phi(\mu).$$

According to Theorem 8.5, $[a, b]$ is the spectrum of the averaging of the function φ over the periodic pseudo-trajectories of the component Ω.

Theorem 8.6. *Let $[a, b]$ be the averaging spectrum of the function φ over the component Ω Then:*

1. $\mathcal{M}(b) = \left\{\mu \in \mathcal{M}(\Omega) : \int\limits_M \varphi d\mu = b\right\}$

   is a non-empty convex compact subset of $\mathcal{M}(\Omega)$;
2. *The extreme points of $\mathcal{M}(b)$ are the ergodic measures μ^* such that $\int\limits_M \varphi d\mu^* = b$;*
3. *for any measure $\mu \in \mathcal{M}(b)$, there exists a partition of the manifold M (modulo null sets) into invariant measurable subsets $\Omega_\alpha, \alpha \in A$, where A is an abstract Lebesgue space with measure β and each Ω_α is equipped with an ergodic invariant measure $\mu_\alpha \in \mathrm{ext}\mathcal{M}(b)$ such that*

$$\int\limits_M \varphi d\mu = \int\limits_A \left(\int\limits_M \varphi d\mu_\alpha\right) d\beta. \tag{8.9}$$

   for any integrable function φ.
4. *The same results are valid for $a = \min\limits_{\mathcal{M}(\Omega)} \Phi(\mu)$.*

Proof

1. According to Theorem 8.5, there exists a measure μ^* such that $b = \int\limits_M \varphi d\mu^*$. Therefore, $\mathcal{M}(b)$ is non-empty. The functional $\Phi(\mu)$ is linear and continuous, so $\mathcal{M}(b) = \Phi^{-1}(b)$ is a convex compact set in the weak topology.
2. By Theorem 6.2.

$$b = \int\limits_M \varphi d\mu^* = \int\limits_A \left(\int\limits_M \varphi d\mu_\alpha\right) d\beta, \tag{8.10}$$

where A is an abstract Lebesgue space with measure β, and there is a division of M (modulo nullsets) into invariant measurable subsets $\Omega_\alpha, \alpha \in A$, and each Ω_α is equipped with an ergodic measure μ_α. We set

$$\psi(\alpha) = \int\limits_M \varphi d\mu_\alpha.$$

Therefore, $\psi(\alpha)$ is a measurable function and $b = \int_A \psi(\alpha) d\beta$. We fix $\varepsilon > 0$ and construct the β-measurable sets $A_1 = \psi^{-1}([a, b-\varepsilon])$ and $A_2 = \psi^{-1}((b-\varepsilon, b])$. The set A can be written as $A = A_1 + A_2$. Arguing by contradiction, we will show that the measure $\beta(A_1) = 0$ is equal to zero. Let us assume that the β-measure of A_1 is $\delta > 0$. Then the β-measure of the set A_2 is $1 - \delta$. We have

$$b = \int_A \psi(\alpha) d\beta = \int_{A_1} \psi(\alpha) d\beta + \int_{A_2} \psi(\alpha) d\beta \le (b-\varepsilon)\delta + b(1-\delta) = b - \varepsilon\delta.$$

If $\delta > 0$, we obtain $b \le b - \varepsilon\delta$, which is a contradiction. Therefore, the β-measure of the set $A_2 = \psi^{-1}((b-\varepsilon, b])$ is 1 for any $\varepsilon > 0$. This gives us a monotonic sequence of nested sets $\{\psi^{-1}((b-\varepsilon, b]), \varepsilon > 0\}$ such that

$$\bigcap_{\varepsilon>0} \psi^{-1}([b-\varepsilon, b]) = \psi^{-1}(b).$$

By monotonicity,

$$\beta\big(\psi^{-1}(b)\big) = \beta\left(\bigcap_{\varepsilon>0} \psi^{-1}((b-\varepsilon, b])\right) = \lim_{\varepsilon\to 0} \beta\left(\bigcap_{\varepsilon>0} \psi^{-1}((b-\varepsilon, b])\right) = 1.$$

The Lebesgue space of A is defined with respect to β-measure up to null-sets. Therefore, we can assume that A coincides with

$$\psi^{-1}(b) = \left\{ \mu_\alpha \in ext(\mathcal{M}(\Omega)) : \int_M \varphi d\mu_\alpha = b \right\}.$$

According to the above, the family of ergodic measures

$$\left\{ \mu_\alpha \in ext(\mathcal{M}(\Omega)) : \int_M \varphi d\mu_\alpha = b \right\}$$

forms the set of extreme points of the convex set $\mathcal{M}(b)$.

3. The third assertion of the theorem follows from the proven Statement 2 and Theorem 6.2. The proof is over.

8.8 Mean Convergence

Definition 8.4 *We say that a sequence* $\omega_n = \{x_n(k), k \in \mathbb{Z}\}$ *of periodic* ε_n-*trajectories converges in the mean as* $\varepsilon_n \to 0$ *if for any continuous function* φ *the mean values over the periods.*

$$\overline{\varphi}(\omega_n) = \frac{1}{p_n}\sum_{k=1}^{p_n}\varphi(x_n(k))$$

converge as $n \to \infty$, *where* p_n *is the period of the pseudo-trajectory* ω_n.

Theorem 8.7 *If a sequence* ω_n *of periodic* ε_n-*trajectories converges in the mean as* $\varepsilon_n \to 0$ *then there exists an invariant measure* μ *such that for any continuous function* φ,

$$\lim_{n\to\infty} \overline{\varphi}(\omega_n) = \int \varphi d\mu.$$

Proof Let us define a functional Φ on the space C^0 of continuous functions by setting

$$\Phi(\varphi) = \lim_{n\to\infty} \overline{\varphi}(\omega_n).$$

The constructed functional is bounded, linear, and positive definite; so according to the Riesz's theorem (see [3]), it can be represented as an integral

$$\Phi(\varphi) = \int \varphi d\mu,$$

where μ is a measure on the manifold M. To show invariance of the constructed measure μ, we need to check the relation:

$$\Phi(\varphi) = \Phi(\varphi(f)),$$

which is equivalent to the relation

$$\lim_{n\to\infty} \overline{\varphi}(\omega_n) = \lim_{n\to\infty} \overline{\varphi}(f(\omega_n)),$$

where $f(\omega_n) = \{y_1, y_2, \cdots, y_p = y_0\}, y_i = f(x_i)$. The sequence $f(\omega_n)$ is periodic. Let us show that $f(\omega_n)$ is an $\eta(\varepsilon)$-trajectory, where $\eta(\cdot)$ is the module of continuity of the mapping f. Since ω_n is a periodic ε-trajectory, then $\rho(f(x_i), x_{i+1}) < \varepsilon$. We have

$$\rho(f(y_i), y_{i+1}) = \rho(f(y_i), f(x_{i+1})) < \eta(\rho(y_i, x_{i+1})) = \eta(\rho(f(x_i), x_{i+1}))) \leq \eta(\varepsilon).$$

Thus, $f(\omega_n)$ is a periodic $\eta(\varepsilon)$-trajectory. Next,

$$|\varphi(x_{i+1}) - \varphi(y_i)| = |\varphi(x_{i+1}) - \varphi(f(x_i))| < \eta(\rho(x_{i+1}, f(x_i)) < \theta(\varepsilon),$$

where $\theta(\cdot)$ is the modulus of continuity of the function φ. We have

$$\begin{aligned}\overline{\varphi}(\omega) &= \frac{1}{p}\sum_{i=1}^{p}\varphi(x_i) = \frac{1}{p}\sum_{i=0}^{p-1}\varphi(x_{i+1})\\ &= \frac{1}{p}\sum_{i=0}^{p-1}\varphi(y_i) + \frac{1}{p}\sum_{i=0}^{p-1}(\varphi(x_{i+1}) - \varphi(y_i))\\ &= \overline{\varphi}(f(\omega_n)) + E,\end{aligned}$$

where $|E| \leq \frac{1}{p}\sum_{i=0}^{p-1}|\varphi(x_{i+1}) - \varphi(y_i)| < \theta(\varepsilon)$.

If $\varepsilon_n \to 0$, then $\theta(\varepsilon) \to 0$ and, therefore,

$$\lim_{n\to\infty} \overline{\varphi}(\omega_n) = \lim_{n\to\infty} \overline{\varphi}(f(\omega_n)).$$

The proof of theorem is complete.

If ω_n is a sequence of periodic ε_n-trajectories, $\varepsilon_n \to 0$ and a point x_n is marked on each pseudo-trajectory ω_n so that $x^* = \lim_{n\to\infty} x_n$, then, according to proposition 3.5, the limit set of sequence ω_n lies in a component Ω of chain-recurrent set, $x^* \in \Omega$. The component Ω is uniquely determined by point x^*. Moreover, sequence ω_n converges uniformly to Ω.

Theorem 8.8 *Let $\omega_n = \{x_n(1), x_n(2), \ldots, x_n(p_n) = x_n(0)\}$ be a sequence of periodic ε_n-trajectories, and let ε_n tend to zero. Then there exists a subsequence ω_{n_k} and an invariant measure μ such that ω_{n_k} converges in the mean, and the average value of any continuous function φ on ω_{n_k} converges to its average value with respect to the measure μ,*

$$\overline{\varphi}(\omega_{n_k}) = \frac{1}{p_{n_k}}\sum_{j=1}^{p_{n_k}}\varphi(x_{n_k j}) \to \int \varphi d\mu$$

as $k \to \infty$, where p_{n_k} is the period of the pseudo-trajectory ω_{n_k}. Moreover, the support of the measure μ lies in the component Ω of the chain-recurrent set.

Proof We assume that the points $x_n(0)$ converge to some point x^* as n $\to \infty$. If this is not true, we can take a subsequence ω_{n_k} for which the points $x_{n_k}(0)$ converge.

Let $C = \{M(i)\}$ be a closed covering of by cells which are polyhedra intersecting by boundary disks. Let G be the symbolic image of the mapping f for the covering C.

According to the Shadowing Theorem 2.1, if ε_n is small enough, the periodic pseudo-trajectory ω_n is traced by a periodical path ζ of period p. By Proposition 5. 4, this path ζ determines a flow

$$m = \left\{ m_{ij} = \frac{k_{ij}}{p} \right\},$$

where k_{ij} is the number of passes of the periodic path ζ through the edge $i \to j$. The flow m generates a measure μ such that the measure of a measurable set A is given by

$$\mu(A) = \sum_i m_i \frac{\nu(A \cap M(i))}{\nu(M(i))},$$

where ν is the Lebesgue measure, $m_i = \sum_j m_{ij}$. The measure of the cell $M(i)$ is calculated as

$$\mu(M(i)) = \sum_j m_{ij} = \sum_j \frac{k_{ij}}{p} = \frac{k_i}{p},$$

where k_i is the number of times the period path ζ passes through the vertex i.

Now, consider a sequence of subdivisions C_k of the covering C with partition diameters d_k converging to 0. Let G_k be their corresponding sequence of symbolic images. Using the technique described above we construct a subsequence of periodic ε_k-trajectories, a sequence of flows m_k on G_k, and a sequence of measures μ_k on M. According to Theorem 5.6 there is a subsequence of measures μ_{k_t} that converges to an invariant measure μ in the weak topology. This means that for every continuous function φ

$$\int \varphi d\mu_{k_t} \to \int \varphi d\mu$$

as $t \to \infty$.

During construction of this sequence, we distinguish two subsequences: the first one ω_{n_k} was obtained when we constructed the sequence of flows m_k on G_k; the second one was extracted from ω_{n_k} to construct a convergent sequence of measures μ_{k_t} on M. In the future, to avoid confusion in the notation, we will assume that all sequences coincide with the original one.

To finish the proof, we need to show that

$$\left| \overline{\varphi}(\omega_n) - \int \varphi d\mu_n \right| \to 0$$

as $n \to \infty$, where $\omega_n = \{x_1, x_2, \cdots x_p = x_0\}$. Taking into account that the boundary disk has zero the Lebesgue measure, we can rewrite

$$\int \varphi d\mu_n = \sum_i \int_{M(i)} \varphi d\mu_n = \sum_i \varphi(x_i)\mu_n(M(i)) = \sum_i \varphi(x_i)\frac{k_i}{p},$$

where x_i is a point in $M(i)$ determined by the mean value theorem. The number k_i coincides with the number of visits of the pseudo-trajectory ω_n to the cell $M(i)$. The average value

$$\overline{\varphi}(\omega_n) = \frac{1}{p}\sum_{j=1}^{p} \varphi(x_j)$$

can be rewritten as follows. We collect all terms corresponding to the points $x_j \in M(i)$ and then sum over i. We obtain

$$\overline{\varphi}(\omega_n) = \frac{1}{p}\sum_i \sum_{x_j \in M(i)} \varphi(x_j).$$

The number of terms in the last sum is the number of times the pseudo-trajectory ω_n passes of through the cell $M(i)$. Then we can write

$$\left|\overline{\varphi}(\omega_n) - \int \varphi d\mu_n\right| \le \frac{1}{p}\sum_i \sum_{x_j \in M(i)} |\varphi(x_j) - \varphi(x_i)| \le \eta(d_n),$$

where x_i and x_j lie in the cell $M(i)$, $\eta(d_n)$ is the modulus of continuity of the function φ, d_n is the diameter of the covering C_n, and $d_n \to 0$. Therefore, the limit of the average over the period coincides with the mean over the measure.

According to the Theorem 3.6, the limit set of the sequence ω_n lies in a component Ω of the chain-recurrent set. The limit point $x^* = \lim_{n\to\infty} x_n(0)$ uniquely defines the component Ω. The sequence ω_n converges uniformly to Ω. The multivalued mapping $h : M \to Ver$ takes each point x to the set of all vertices i such that $x \in M(i)$. The image $h(x)$ of a chain-recurrent point consists of recurrent vertices, and there is a unique class $H(\Omega)$ of equivalent recurrent vertices such that $h(\Omega) \subset H(\Omega)$. According to Theorem 3.9 on localization of the chain-recurrent set, the union $U = \{\cup M(i), i \in H(\Omega)\}$ is a closed neighbourhood of the component Ω. If the covering diameter d tends to zero, then U converges to Ω. By construction, the support of the measure μ_k lies in $U_k = \{\cup M(i), i \in H(\Omega), M(i) \in C_k\}$. Therefore, the support of the measure μ lies in the component $\Omega = \bigcap_k U_k$. Proof is complete.

Theorem 8.7 states that a sequence of periodic pseudo-trajectories determines some invariant measure. The converse is stated by the next theorem: an invariant measure generates a sequence of periodic pseudo-trajectories that converge in the mean to the average over a given measure.

Theorem 8.9 *If the support of the invariant measure μ lies in a single component of a chain-recurrent set, then there exists a sequence of periodic ε_n-trajectories ω_n, $\varepsilon_n \to 0$ such that for any continuous function, φ the average value over a period $\overline{\varphi}(\omega_n)$ converges to the average with respect to that measure.*

Proof Let Ω be a component of the chain-recurrent set, and let μ be an invariant measure with support in Ω. Our goal is to prove that there exists a sequence of periodic pseudo-trajectories that converge in the mean to averaging with respect to the invariant measure. The proof of this statement is similar to the proof of Theorem 8.5.

Accordingly, this proof will be given in abbreviated form. Consider a covering $C = \{M(i)\}$ of the manifold M whose cells are closed polyheda intersecting in boundary disks. Let d be the diameter of this covering. Let G be a symbolic image of the mapping f with respect to the covering C, and let φ be a continuous function defined on M. For each vertex i, we define the number $b[i]$ as follows. We fix a point x_i in each cell $M(i)$ and set $b[i] = \varphi(x_i)$. It is clear that such marking depends on our choice of $x_i \in M(i)$. If x_i^* is another point in $M(i)$, then

$$\left|\varphi(x_i) - \varphi\left(x_i^*\right)\right| \le \eta(d),$$

where $\eta(d)$ is the continuity modulus of the function φ. Using the covering C, we construct a measurable partition $C^* = \{M^*(i)\}$ consisting of polyhedra, in which every boundary disk belongs to only one of neighboring cell. The invariant measure μ generates a flow m on G according to the equation

$$m_{ij} = \mu\Big(f\left(M^*(i)\right)\bigcap M^*(j)\Big) = \mu(M^*(i)\bigcap f^{-1}\left(M^*(j)\right),$$

Let $H(\Omega) = H$ be the unique class of equivalent recurrent vertices corresponding to Ω. Then,

$$U = \{\cup M(i), \quad i \in H(\Omega)\}$$

is a closed neighborhood of the component Ω and U converges to Ω as $d \to 0$. Since the support $supp(\mu)$ lies in Ω, the flow m lies in $H(\Omega)$. According to Theorem 5. 2, near the flow m there exists a periodic flow m^* such that the distance $\rho(m, m^*)$ is sufficiently small. This periodic flow m^* is determined by the periodic path ω^* on G. According to Theorem 2.1, this path generates the desired periodic pseudotrajectory ω.

The mean value over the flow m_{ij} is the number

$$\sum_{ij} m_{ij} b[i] = \sum_i m_i b[i] = \sum_i \mu\left(M^*(i)\right)\varphi(x_i).$$

According to the mean value theorem, we have

$$\int_M \varphi(x)d\mu = \sum_i \int_{M^*(i)} \varphi(x)d\mu = \sum_i \varphi(x_i^*)\mu(M^*(i)),$$

where the middle point x_i^* lies in the cell $M(i)$. As a result, we obtain the inequality

$$\left|\sum_i m_i b[i] - \int_M \varphi(x)d\mu\right| \leq \theta(d).$$

For every $\delta > 0$ there exists a periodic flow $m^* = \left\{m_{ij}^*\right\}$ such that $\sum_{ij} |m_{ij} - m_{ij}^*| < \delta$. The flow m^* is generated by the periodic path $\omega^* = \{i_k, k = 1, 2, \cdots, p\}$ such that $m_{ij}^* = \frac{k_{ij}}{p}$, where k_{ij} is the number of times the path ω^* passes through the edge $i \to j$. We have

$$\left|\sum_i m_i b[i] - \sum_i m_i^* b[i]\right| = \left|\sum_i b[i] \sum_j (m_{ij} - m_{ij}^*)\right| \leq K\delta,$$

where $K = max|\varphi|$. The averaging over the periodic flow m^* coincides with averaging over the periodic path ω^*

$$b(\omega^*) = \frac{1}{p}\sum_{ij} k_{ij} b[i] = \frac{1}{p}\sum_k \varphi(x_{i(k)}).$$

By Theorem 2.1, the periodic path ω^* generates the periodic pseudo-trajectory $\omega = \{x_k, k = 1, 2, \cdots, p\}$, such that $x_k \in M(i_k)$ and ω is a $\frac{3}{2}d$-trajectory. Since $x_{i(k)}$ and x_k both lie in $M(i_k)$, we have

$$\left|\frac{1}{p}\sum_k \varphi(x_{i(k)}) - \frac{1}{p}\sum_k \varphi(x_{i_k})\right| < \eta(d).$$

Summing up the obtained inequalities, we get

$$\left|\int_M \varphi d\mu - b(\omega)\right| \leq 2\eta(d) + K\delta.$$

Let ε be an arbitrary positive number, then we can choose $d > 0$ and $\delta > 0$ such that $2\eta(d) + K\delta < \varepsilon$. In this case, the difference between the average of the function

φ over the periodic pseudotrajectory ω and the average of this function over the measure μ is less than ε.

Finally, consider a sequence of coverings C_k with diameters d_k converging to zero. This procedure allows us to find a sequence of periodic ε_n-trajectories $\{\omega_n, \varepsilon_n \to 0\}$ that their means $\overline{\varphi}(\omega_n)$ converge to the averaging $\int\limits_M \varphi d\mu$ with respect to the measure. Theorem is proved.

8.9 Recurrence and Periodicity

Let $\omega_n = \{x_n(1), x_n(2), \cdots, x_n(p_n) = x_n(0)\}$ be a sequence of periodic ε_n-trajectories, $\varepsilon_n \to 0$. On each pseudo-trajectory ω_n, we mark a point x_n such that $x^* = \lim\limits_{n\to\infty} x_n$. The limit point x^* lies in a component Ω of the chain-recurrent set. Then the sequence ω_n uniformly converges to the component Ω (according to the Theorem 3.6), and the pseudo-trajectories ω_n converge pointwise to the trajectory $Tr(x^*)$. A sequence of periodic $\mathrm{\varepsilon_n}$-trajectories, $\mathrm{\varepsilon_n} \to 0$, converges uniformly to a trajectory $\mathrm{Tr} = \left\{\mathrm{y\ (k)} = \mathrm{f^k(y(0))}\right\}$, if $\mathrm{sup_k(\rho(x_n(k), y(k))}$ converges to zero as $\mathrm{n} \to \infty$.

Theorem 8.10 *If a sequence ω_n of periodic ε_n-trajectories, $\varepsilon_n \to 0$, converges uniformly to a trajectory Tr, then this trajectory is recurrent. The limit set of this sequence coincides with the closure of the trajectory Tr and is a minimum set.*

Proof The pseudo-trajectory $\omega_n = \{x_n(k), k \in \mathbb{Z}\}$ has period p_n. Let us fix n and set $r_n = sup_k \rho(x_n(k), y(k))$. Then, according to the uniform convergence assumption, we have $r_n \to 0$ as $n \to \infty$. Let B(x, r) be the ball with radius r and center x. Since the pseudo-trajectory ω_n is p_n-periodic, then the union of p_n balls of the form.

$$U_n = \left\{ \bigcup_{k=1}^{p_n} B(x_n(k), r_n), x_n(k) \in \omega_n \right\}$$

contains the entire trajectory $Tr = \{y(k), k \in \mathbb{Z}\}$. In this case, $x_n(k) = x_n(k + zp_n), z \in Z$, and the distance

$$\rho(y(k + zp_n), x_n(k)) = \rho(y(k + zp_n), x_n(k + zp_n)) \le r_n$$

for $z \in Z$. Therefore, the distance $\rho(y(k + zp_n), y(k)) \le 2r_n$ for $z \in Z$, and a ball of radius $2r_n$ centered at $y(k)$ contains points of the form $y(k + zp_n), z \in Z$. Thus, the union of p_n balls of the form

$$U_n^*(k_0) = \left\{ \bigcup_{z} B(y(k_0 + z), 2r_n), 0 < z \le p_n \right\}$$

contains the entire trajectory Tr for any number $k_0 \in Z$. Let us show that the trajectory Tr is recurrent. We fix $\varepsilon > 0$. Since $r_n \to 0$, we can find $r_n < 1/2\varepsilon$, which determines the number n and the period p_n of the pseudo-trajectory ω_n. From the above construction, it follows that the $2r_n$-neighbourhood of any segment of length p_n of the trajectory Tr contains the entire trajectory Tr. Therefore, Tr is a recurrent trajectory. Since $r_n \to 0$ as $n \to \infty$, we see that the limit set of the sequence ω_n coincides with the closure of the trajectory Tr. The Birkhoff theorem (see [3]) states that the closure of a recurrent trajectory is a compact minimal set. The proof of the theorem is complete.

Suppose that the half-trajectory $T = \{x(k) = f^k(x(0)), k \in Z^+\}$ has a limit point x^*, that is, there is a sequence $x(k_m) \to x^*$ for $m \to \infty$. Thus, for any $\varepsilon > 0$, there are points $x(k_1), x(k_2)$ such that the distance between them is $\rho(x(k_1), x(k_2)) < \varepsilon, k_1 \neq k_2$. Let $k_1 < k_2$, then $\rho(f^{k_2-k_1}(f^{k_1}(x(0)), f^{k_1}(x(0))) < \varepsilon$. Therefore,

$$\omega = \left\{y(0) = f^{k_1}(x(0)), y(1) = f^{k_1+1}(x(0)), \cdots, y(p-1) = f^{k_2-1}(x(0)), y(p) = y(0)\right\}$$

is a p-periodic ε-trajectory. In this way, we construct a sequence ω_n of periodic ε_n-trajectories, $\varepsilon_n \to 0$. On each pseudotrajectory ω_n, the points $y_n = f^{k_n}(x(0))$ are marked, and $y_n = f^{k_n}(x(0)) \to x^*$ as $n \to \infty$. According to the Theorem 8.8, there is a subsequence ω_{n_m} and an invariant measure μ such that the mean value over ω_{n_m} of any continuous function converge to its average over the measure μ,

$$\overline{\varphi}(\omega_{n_m}) = \frac{1}{p_m}\sum_{k=0}^{p_m-1} \varphi(y(k)) \to \int_M \varphi d\mu$$

as $m \to \infty$. The support of μ lies in ω-limit set of the trajectory T.

Now let be recurrent trajectory $Tr = \{x(k) = f^k(x(0)), k \in \mathbb{Z}\}$. In this case, the marked and limit points coincide with the starting point

$$y_n = x(0) = x^*, k_1 = 0 \text{ and } y(k) = f^k(x^*).$$

It follows that recurrent trajectory Tr has a sequence of integers $p_n \to \infty$ and an invariant measure μ such that

$$\lim_{n\to\infty} \frac{1}{p_n}\sum_{k=o}^{p_n-1} \varphi(f^k(x(0))) = \int_M \varphi d\mu. \tag{8.11}$$

Moreover, the support of the measure μ coincides with the closure $\overline{Tr} = \Omega$ of the recurrent trajectory, which is a minimal set. Question arises how invariant measure is related to the subsequence $p_m \to \infty$?

We need results from [4]. A point $x \in \Omega$ is called quasiregular if for any continuous function φ there is a limit

$$\lim_{n\to\infty} \frac{1}{n} \sum_{k=0}^{n-1} \varphi\big(f^k(x)\big). \tag{8.12}$$

An example of a minimal set with an irregular point is given in [4]. This means that the limit of the subsequence (8.11) always exists, but the general limit (8.12) does not necessary exist. The following theorem holds.

Theorem 8.11 [4] *Every minimal set is either an ergodic set or contains at least one non-quasiregular point, that is, a point $x \in \Omega$ for which there exists a continuous function φ such that the limit*

$$\overline{\varphi}(x) = \lim_{n\to\infty} \frac{1}{n} \sum_{k=1}^{n} \varphi\big(f^k(x)\big)$$

does not exist.

In the latter case, the closure $\overline{Tr} = \Omega$ of the recurrent trajectory is the support for more than one ergodic measure. The following theorem is proven in [4].

Theorem 8.12 [4] *For a compact dynamical system (Ω, f) the following conditions are equivalent at a point $p \in \Omega$.*

(1) *for each continuous function φ, the sequence*

$$\frac{1}{m} \sum_{k=0}^{m-1} \varphi\big(f^{k+s}(p)\big)$$

converges uniformly in $s \in Z$ as $m \to \infty$;

(2) *the subsystem $\big(\overline{Tr}, f\big)$ is strictly ergodic (allows a unique invariant measure).*

Theorem 8.13 *If a sequence ω_n of periodic ε_n-trajectories converges in mean as $\varepsilon_n \to 0$ and uniformly to a trajectory Tr as $n \to \infty$, then the closure of this trajectory is a minimal strictly ergodic set.*

Proof Theorem 8.10 states that, under our conditions, the trajectory Tr is recurrent and its closure $\overline{Tr} = \Omega$ is a minimal set. It follows from Theorem 8.12 that to complete the proof, it is necessary to prove that the sequence.

$$\frac{1}{m} \sum_{k=0}^{m-1} \varphi\big(f^{k+s}(p)\big)$$

converges uniformly in $s \in Z$ as $m \to \infty$, where $p \in Tr$.

Let $\omega_n = \{x_n(k), k \in Z\}$ be a sequence of periodic ε_n-trajectories, $\varepsilon_n \to 0$, and let p_n be the period of ω_n. The average value of the function φ is

$$\overline{\varphi}(\omega_n) = \lim_{m\to\infty} \frac{1}{m}\sum_{k=0}^{m-1} \varphi(x_n(k)) = \frac{1}{p_n}\sum_{k=0}^{p_n-1} \varphi(x_n(k)).$$

We assume without loss of generality that $x_n(0)$ converges to some point p. Then $y(0) = p$, $Tr = \{y(k) = f^k(p), k \in Z\}$, and $r_n = sup_k \rho(x_n(k), y(k))$ converges to zero as $n \to \infty$. The difference between the average values satisfies the relation

$$\left|\frac{1}{m}\sum_{k=0}^{m-1}\varphi\big(f^{k+s}(p)\big) - \frac{1}{m}\sum_{k=0}^{m-1}\varphi(x_n(k+s))\right| \le$$

$$\frac{1}{m}\sum_{k=0}^{m-1}\left|\varphi\big(f^{k+s}(p)\big) - \varphi(x_n(k+s))\right| \le \theta(r_n),$$

where $\theta(r)$ is the modulus of continuity of the function φ, and numbers m and s are arbitrary. By assumption, the sequence ω_n converges in mean, so

$$\lim_{n\to\infty} \overline{\varphi}(\omega_n) = \int_M \varphi d\mu,$$

where μ is an invariant measure. Let us show that

$$\lim_{m\to\infty}\frac{1}{m}\sum_{k=0}^{m-1}\varphi\big(f^{k+s}(p)\big) = \int_M \varphi d\mu$$

and this convergence is uniform in s. We have

$$\left|\frac{1}{m}\sum_{k=0}^{m-1}\varphi\big(f^{k+s}(p)\big) - \int_M \varphi d\mu\right| \le \left|\frac{1}{m}\sum_{k=0}^{m-1}\varphi\big(f^{k+s}(p)\big) - \frac{1}{m}\sum_{k=0}^{m-1}\varphi(x_n(k+s))\right| +$$

$$\left|\frac{1}{m}\sum_{k=0}^{m-1}\varphi(x_n(k+s)) - \overline{\varphi}(\omega_n)\right| + \left|\overline{\varphi}(\omega_n) - \int_M \varphi d\mu\right| \le$$

$$\theta(r_n) + \left|\frac{1}{m}\sum_{k=0}^{m-1}\varphi(x_n(k+s)) - \frac{1}{p_n}\sum_{k=0}^{p_{n-1}}\varphi(x_n(k))\right| + \delta_n,$$

where $\delta_n \to 0$ as $n \to \infty$. Choosing an appropriate n, we can make both the first and last terms small enough for all m and s. Let m be large enough so that the quantity

$$\left|\frac{1}{m}\sum_{k=0}^{m-1}\varphi(x_n(k+s)) - \frac{1}{p_n}\sum_{k=0}^{p_{n-1}}\varphi(x_n(k))\right|$$

is sufficiently small uniformly in s for fixed n. Since the sequence ω_n is periodic with period p_n, we have

$$\sum_{k=0}^{p_{n-1}}\varphi(x_n(k+s)) = p_n\overline{\varphi}(\omega_n)$$

for any s. Let $m = tp_n + r$, where t is the integer part of the fraction m/p_n, and the remainder r satisfies the inequality $0 \le r < p_n$. Then

$$\frac{1}{m}\sum_{k=0}^{m-1}\varphi(x_n(k+s)) = \frac{1}{m}(\sum_{k=0}^{p_n-1}\varphi(x_n(k+s)) + \sum_{k=0}^{p_n-1}\varphi(x_n(k+p_n+s)) + \cdots$$

$$+\sum_{k=0}^{p_n-1}\varphi(x_n(k+(t-1)p_n+s)) + \sum_{k=0}^{r-1}\varphi(x_n(k+tp_n+s)))$$

$$= \frac{1}{m}\left(tp_n\overline{\varphi}(\omega_n) + \sum_{k=0}^{r-1}\varphi(x_n(k+tp_n+s))\right) = \overline{\varphi}(\omega_n) + \alpha(t),$$

where

$$\alpha(t) = \frac{-r}{tp_n+r}\overline{\varphi}(\omega_n) + \frac{1}{tp_n+r}\sum_{k=0}^{r-1}\varphi(x_n(k+tp_n+s)).$$

Taking into account the boundedness of the function φ, $|\varphi| < C$ and the inequality $r < p_n$, we obtain the estimate $|\alpha(t)| \le {}^{2C}/_{(t-1)}$. Thus, for any $\varepsilon > 0$, we can choose n so that $\theta(r_n) + \delta_n \le 2\varepsilon/3$. Then choose t^* so that ${}^{2C}/_{(t^*-1)} \le \varepsilon/3$. Then

$$\left|\frac{1}{m}\sum_{k=0}^{m-1}\varphi\big(f^{k+s}(p)\big) - \int_M \varphi d\mu\right| \le \varepsilon$$

for $m > t^* p_n$, uniformly in s. In this case, according to the Theorem 8.10, the limit set of the sequence ω_n is that same as the closure of the recurrent trajectory Tr and is a minimal set with unique ergodic measure μ. Theorem is proved.

8.10 Periodicity and Ergodicity

Theorem 8.14 *Consider a sequence $C_t = \{M_t(i)\}$ of subdivisions. Let the diameters d_t of the coverings C_t converge to zero, and let G_t be a corresponding sequence of symbolic images. Suppose that a periodic path $\omega_t = \{i_t(1), i_t(2), \cdots, i_t(p_t) = i_t(0)\}$ of period p_t is given on each G_t, and a sequence of paths ω_t is consistent. Then the following statements are true.*

(1) *There exists a recurrent trajectory $T = \{x_k : x_{k+1} = f(x_k),\}$, for which $x_k \in M\left(i_{t(k)}\right)$, for any t.*
(2) *A sequence of periodic pseudo-trajectories*

$$T_t = \left\{x_t(k) \in M_t\left(i_{t(k)}\right), k \in \mathbb{Z}\right\}$$

(which are traces of the paths ω_t) uniformly converges to a trajectory T such that $sup_k \rho(x_t(k), x_k) \leq d_t$.

(3) *The sequence of sets $P_t(\omega_t) = \bigcup_k M_t(i_t(k))$ is nested: $P_{t+1} \subset P_t$, and the limit $\lim\limits_{t\to\infty} P_t = \bigcap_t P_t$ coincides with the closure of the trajectory T.*
(4) *The sequence of periodic pseudo-trajectories T_t converges in mean, and an invariant measure μ exists such that, for any continuous function φ, the average values $\overline{\varphi}(\omega_t)$ converge to $\int\limits_M \varphi d\mu$ as $t \to \infty$.*
(5) *The closure of the trajectory T is a minimal strictly ergodic set of the measure μ and its support*

$$supp\mu = \lim_{t\to\infty} \bigcup\nolimits_k M_t(i_t(k)).$$

(6) *If $\{x_t(n) \in M_t(i_t(k)), 1 \leq n \leq p_t\}$ is a trace of the path ω_t, then. the discrete measure*

$$\mu_t^* = \frac{1}{p_t} \sum_{n=1}^{p_t} \delta(x_t(n))$$

converges weakly to the ergodic measure μ as $t \to \infty$, where $\delta(x)$ is the Dirac delta-function.

Proof Under the conditions of the theorem, there is a sequence of subdivisions $C_t = \{M_t(i)\}$, a sequence of the symbolic images G_t connected by the mapping $s : G_{t+1} \to G_t$, a sequence of the flow spaces $\mathcal{M}(G_t)$ connected by the mapping $s^* : \mathcal{M}(G_{t+1}) \to \mathcal{M}(G_t)$. Each periodic path $\omega_t = \{i_t(1), i_t(2), \cdots, i_t(p_t) = i_t(0)\}$ of period p_t lies on the symbolic image G_t and $s(\omega_{t+1}) = \omega_t$. Statements 1–2 are consequences of Theorem 3.1. Let us briefly recall their proofs.

Proof of Statement 1 Let k be fixed and consider a sequence of cells $\{M_t(i_t(k)), t = 1, 2, ...\}$ from the sequence of subdivisions C_t. The consistency of periodic paths ω_t implies that $s(i_{t+1}(k)) = i_t(k)$. This means that the cell $M_{t+1}(i_{t+1}(k))$ is included in the subdivision of the cell $M_t(i_t(k))$. In this case, there are inclusions

$$M_1(i_1(k)) \supset M_2(i_2(k)) \supset \cdots \supset M_t(i_t(k)) \supset M_{t+1}(i_{t+1}(k)) \supset \cdots . \quad (8.13)$$

Since the cells are closed, and their diameters tend to zero together with d_t, there is a single point

$$x_k = \lim_{t\to\infty} M_t(i_t(k)) = \bigcap_t M_t(i_t(k)).$$

The sequence $\{x_k : x_{k+1} = f(x_k), k \in \mathbb{Z}\}$ is a trajectory, see the proof of the Statement 1 of the Theorem 3.1 for details. Note that for each t the entire trajectory $T = \{x_k\}$ lies in the union of the cells of the periodic path $\omega_t = \{i_t(1), i_t(2), \cdots, i_t(p_t) = i_t(0)\}$,

$$T \subset P(\omega_t) = \bigcup_k M\_t(i\_t(k)).$$

This is enough to prove the recurrence of the trajectory T; for details, see the proof of the Statement 4 of Theorem 3.1.

Proof of Statement 2 Let $\omega_t = \{i_t(k), 1 \le k \le p_t\}$ be a consistent sequence of periodic paths on $\{G_t\}$. Fixing t, we define a periodic pseudotrajectory $T_t = \{x_t(k) \in M_t(i_t(k)), 1 \le k \le p_t\}$ as a trace of the path ω_t. According to the Theorem 2.1, T_t is an ε-trajectory for any $\varepsilon > d_t + \theta(d_t)$, where θ is the modulus of continuity of the mapping f. According to the construction, the point $x_t(k)$ of the pseudotrajectory T_t and the point x_k of the trajectory T lie in the same cell $M_t(i_t(k))$ for each k and t. Then the distance between these points does not exceed the diameter d_t of the covering C_t. Consequently, the sequence of pseudo-trajectories T_t converges uniformly to a recurrent trajectory $T = \{x_k\}$ such that the distance $\rho(x(k), x_t(k)) < d_t$.

Statement 3 follows from the statement 3 of Theorem 3.1.

Proof of Statement 4 If the symbolic image G has a periodic path ω of period N, then there is a flow m on G such that $m_{ij} = k_{ij}/N$, where k_{ij} is the number of passes of ω through the arc $i \to j$. Thus, the consistent sequence $\omega_t = \{i_t(k)\}$ of periodic paths generates a consistent sequence of periodic flows $m(\omega_t)$. In each cell $M_t(i)$, we select a point $x_t(i)$. The flow $m(\omega_t)$ generates the measure μ_t on the manifold M according to the formula

$$\mu_t = \sum_i m_t(i)\delta(x_t(i)), \; m_t(i) = \sum_j m_t(ij) = \sum_j \frac{k_t(ij)}{p_t} = \frac{k_t(i)}{p_t},$$

where $\delta(x)$ is the Dirac δ-function concentrated at x, where $k_t(ij)$ is the number of passes of path ω_t through the arc $i \to j$, $k_t(i)$ is the number of passes of path ω_t through vertex i. The constructed measure concentrates at the points $x_t(i)$. For any continuous function φ, we have

$$\int \varphi d\mu_t = \sum_i m_t(i)\varphi(x_t(i)) = \sum_i \frac{k_t(i)}{p_t}\varphi(x_t(i)). \tag{8.14}$$

According to Theorem 5.10, the sequence of discrete measures μ_t converges to an invariant measure μ in the weak topology as $d_t \to 0$. To construct measure μ_t, it is necessary to select the point in each cell $M_t(i)$. The path ω_t generates the periodic pseudotrajectory

$$T_t = \{x_t(k) \in M_t(i_t(k)), 0 \leq k \leq p_t\}.$$

We select a point in $M_t(i_t(k))$ such that $i_t(i) = i_t(k)$. According to Theorem 2.1, a periodic sequence of points

$$T_t = \{x_t(k) \in M_t(i_t(k)), 0 \leq k \leq p_t\}$$

is the trace of the path ω_t. Each point $x_t(k)$ lies in $M_t(i_t(k))$ and it is possible that $i_t(k_1) = i_t(k_2)$ but that $x_t(k_1) \neq x_t(k_2)$. In this case, the points $x_t(k_1)$ and $x_t(k_2)$ lie in the same cell, that is, distance $\rho(x_t(k_1), x_t(k_2)) \leq d_t$. Our goal is to show that for any continuous function mean value

$$\overline{\varphi}(T_t) = \frac{1}{p_t}\sum_{k=1}^{p_t}\varphi(x_t(k))$$

converges as $t \to \infty$. Let's compare the resulting average with the formula (8.14). Note that path ω_t passes through vertices i when pseudo-trajectory T_t passes through cells $M_t(i)$. It follows that $k_t(i) \neq 0$ in (8.14) only for vertices of the periodic path $\omega_t = \{i_t(k), 1 \leq k \leq p_t\}$. Moreover, $\sum_i k_t(i) = p_t$. Substituting in (8.14)

$$k_t(i)\varphi(x_t(k)) = \sum_{i_t(k)=i} \varphi(x_t(i_t(k))),$$

we obtain

$$\int \varphi d\mu_t = \frac{1}{p_t}\sum_{k=1}^{p_t}\varphi(x_t(i_t(k))).$$

Let us show that average value $\overline{\varphi}(T_t)$ of the function φ over periodic pseudo-trajectory T_t and integral $\int \varphi d\mu_t$ have common limit as $t \to \infty$.

According to the construction, points $x_t(k)$ and $x_t(i_t(k))$ lie in the same cell $M_t(i_t(k))$ and the distance $\rho(x_t(k), x_t(i_t(k))) \leq d_t$. Then

$$\left| \int \varphi d\mu_t - \overline{\varphi}(T_t) \right| \leq \frac{1}{p_t} \sum_{k=1}^{p_t} |\varphi(x_t(i_t(k))) - \varphi(x_t(k))| \leq \eta(d_t) \to 0,$$

where η is the modulus of continuity of the function φ and diameter of the partition $d_t \to 0$ as $t \to \infty$. This means that average $\overline{\varphi}(T_t)$ and integral $\int \varphi d\mu_t$ converge to a common limit. Since the sequence of measures μ_t converges to the invariant measure μ in the weak topology, then

$$\lim_{t\to\infty} \overline{\varphi}(T_t) = \lim_{t\to\infty} \int \varphi d\mu_t = \int \varphi d\mu,$$

that is, the sequence of periodic pseudo-trajectories T_t converges in mean.

Proof of Statement 5 According to Statement 4, the sequence T_t of periodic pseudo-trajectories converges in mean. According to Statement 2, the sequence T_t converges uniformly to the trajectory T. By Theorem 8.13, the closure of this trajectory is the minimal strictly ergodic set and hence the invariant measure μ is ergodic. The closure of the trajectory T is the support of this measure. According to Statement 3, support will be

$$supp\mu = \lim_{t\to\infty} P_t = \bigcap_t \left(\bigcup_k M_t(i_t(k)) \right).$$

Proof of Statement 6 Let $T_t = \{x_t(n) \in M_t(i_t(n)), 1 \leq n \leq p_t\}$ be the trace of the periodic path ω_t and let the discrete measure μ_t have the form

$$\mu_t = \frac{1}{p_t} \sum_{n=1}^{p_t} \delta(x_t(n)).$$

Then

$$\int \varphi d\mu_t = \frac{1}{p_t} \sum_{n=1}^{p_t} \varphi(x_t(n)) = \overline{\varphi}(T_t).$$

According to the proof of Statement 4, the mean values $\overline{\varphi}(T_t)$ converge to the integral $\int \varphi d\mu$ as $t \to \infty$. It follows that the discrete measure μ_t convergences to an ergodic measure μ in the weak topology. Theorem is proven.

Corollary. Statement 6 of this proven theorem allows us to construct a numerical approximation of the ergodic measure μ.

8.11 Examples: Averaging on Circle

1. Consider a continuous dynamical system generated by the differential equation $\varphi' = cos\varphi$ on the unit circle in the plane R^2, where φ is an angle measured from the positive direction of the OX axis counterclockwise. This system has two equilibrium states $\varphi = \pm\frac{\pi}{2}$. The point N corresponds to the equilibrium state $\varphi = \frac{\pi}{2}$, and the point S corresponds to another equilibrium state $\varphi = -\frac{\pi}{2}$. See Fig. 8.1.

So, in the Cartesian coordinate system $N = (0, 1)$, $S = (0, -1)$. The equilibrium state N is asymptotically stable, and S is the unstable equilibrium state, the remaining trajectories go from S to N. The discrete dynamical system is generated by the unit time shift along the trajectories of the continuous system. The system has two invariant delta measures δ(S) and δ(N) concentrated in S and N, respectively. The set $\mathcal{M}$ of all invariant measures is the linear span of $\delta(S)$ and $\delta(N)$. A chain-recurrent set consists of equilibrium states S and N, which form its two components. According to Theorem 8.5, the averaging spectrum of a function $g(\varphi)$ over pseudo-trajectories gives two values of the function at points S and N, $\Sigma = \{g(\pi/2), g(-\pi/2)\}$. For example, averaging spectrum of the ordinate y(φ) gives two values 1 and -1. The set of all invariant measures $\mathcal{M} = \{\alpha\delta(S) + \beta\delta(N), \alpha \geq 0, \beta \geq 0, \alpha + \beta = 1\}$. Averaging y(φ) over all invariant measures $\mathcal{M}$ gives the segment $[-1, 1]$. This means that averaging over pseudo-trajectories of the system does not generate all invariant measures.

2. Consider the system generated by the differential equation $\varphi' = cos^2\varphi$ on the unit circle. A shift per unit of time generates a discrete dynamical system. The system has two fixed points S and N, but these fixed points are "passing" (see Fig. 8.2). The derivative $\varphi' > 0$ for $\varphi \neq \pm\pi/2$, therefore φ(t) increases and the trajectories go counterclockwise from S to N and from N

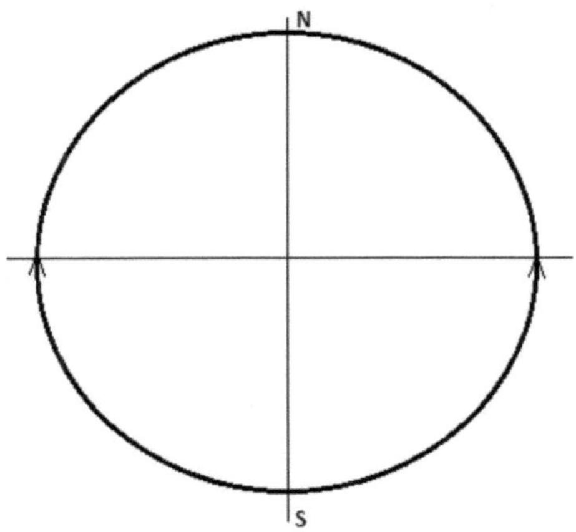

Fig. 8.1 $\varphi' = cos\varphi$

Fig. 8.2 $\varphi' = cos^2\varphi$

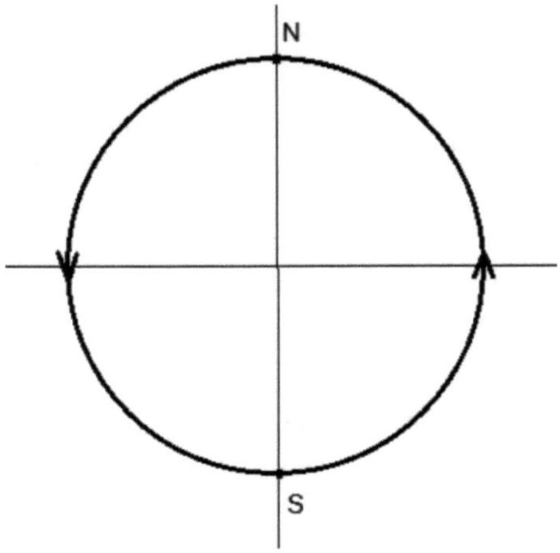

back to S. Invariant measures will remain the same, as in the previous example, $\mathcal{M} = \{\alpha\delta(S) + \beta\delta(N), \alpha \geq 0, \beta \geq 0, \alpha + \beta = 1\}$.

However, a chain-recurrent set has a component that coincides with the unit circle. According to Theorem 8.5, the averaging spectrum of the function y(φ) is the interval [−1, 1], which is obtained by averaging the function y(φ) over these measures $\mathcal{M}$.

3. Finally, we consider the system generated by the differential equation $\varphi' = cos^2(\varphi/2)$ (see Fig. 8.3). The system has only one equilibrium state $\varphi = \pi$ and one invariant measure, which is concentrated in $\varphi = \pi$, that is, $\mathcal{M} = \{\delta(\pi)\}$.

A chain-recurrent set has one component coinciding with the unit circle. According to Theorem 8.5, the average spectrum of the function $y(\varphi)$ consists of a single value 0, which is obtained by averaging the function $y(\varphi)$ over $\delta(\pi)$.

Fig. 8.3 $\varphi' = cos^2(\varphi/2)$

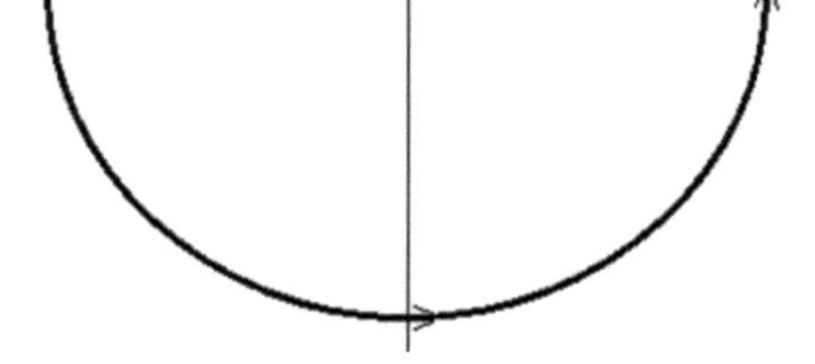

8.12 Volume Exponential

Consider a diffeomorphism $f : M \to M$ on a smooth manifold. The differential $Df = (\partial f/\partial x)$ acts on the tangent space TM from T_xM to $T_{f(x)}M$. The modulus $|det\, Df|$ of the differential determinant is the coefficient of volume change when displaying f. The equality

$$\left|det\, Df^n(x)\right| = \prod_{k=1}^{n} \left|det\, Df\left(f^{k-1}(x)\right)\right|.$$

follows from the recurrence formula

$$\left|det\, Df^n(x)\right| = \left|det\, Df\left(f^{n-1}(x)\right)\right|\left|det\, Df^{n-1}(x)\right|.$$

Meaning

$$\frac{1}{n} ln\left|det\, Df^n(x)\right| = \frac{1}{n}\sum_{k=0}^{n-1} ln|det\, Df(x_k)|,\, x_k = f(x_{k-1}), \quad x_0 = x,$$

is an exponent of volume change along a segment of the trajectory. This exponent is the arithmetic mean of the function $\varphi = ln|det\, Df(x)|$ over $x_0, x_1, \cdots, x_{n-1}$. This observation leads us to following definition.

Let the sequence $\{x_k\}$ be a pseudotrajectory of diffeomorphism f, the value

$$\lambda = \frac{1}{n}\sum_{k=0}^{n-1} ln|det\, Df(x_k)|$$

is an exponent of the change in volume along the segment of pseudo-trajectory. The volume exponent spectrum is called the averaging spectrum Σ of function $\varphi = ln|det\, Df(x)|$ over pseudotrajectories of diffeomorphism f.

According to Theorem 8.5, the spectrum of volume changes consists of segments $[a_k, b_k]$, each segment is generated by a chain-recurrent component Ω_k and coincides with the set of averaging functions $ln|det\, Df(x)|$ over invariant measures focused on this component:

$$\Sigma(\Omega_k) = \left\{\lambda = \int_M ln|det\, Df(x)|d\mu,\, \mu \in \mathcal{M}(\Omega_k)\right\},$$

where $\{\Omega_k\}$ is the complete family of chain-recurrent components.

Proposition 8.10 *Let G be a symbolic image of a dynamical system* $x_{n+1} = f(x_n)$ *with respect to a covering C, and let* Ω *be a chain-recurrent component. Let* $H(\Omega)$ *be the set of equivalent recurrent vertices such that* $\Omega \subset \{\cup M(i), i \in H(\Omega)\}$ *and the interval [a, b] is the extended spectrum* $\Sigma(H)$ *of class* $H(\Omega\}$. *There are numbers* K_* *and* K^* *such that if* $\chi = \{x_k = f^k(x)\}$ *is a trajectory lying in* Ω, *then the inequalities.*

$$K_* exp(pa) \leq |detDf^p(x)| \leq K^* exp(pb).$$

hold for any p.

In particular, if b < 0, then the dynamic system compresses the volume on Ω.

Similar estimates will be proven in Sect. 9.2, and therefore the proof is skipped here.

References

1. Osipenko, G.S., Romanovsky, J.V., Ampilova, N.B., Petrenko, E.I.: Computation of Morse spectrum. J. Math. Sci. **120**(2), 1155–1166 (2004)
2. Georgiadis, L., Goldberg, A.V., Tarjan, R.E., Werneck, R.F.: An experimental study of minimum mean cycle algorithms. In: Proceedings 6th International Workshop on Algorithm Engineering and Experiments, ALENEX, pp. 1–13 (2009)
3. Nemytski, V.V., Stepanov, V.V.: Qualitative Theory of Dynamical Systems. Princeton University Press. Russian original 1949 (1960)
4. Oxtoby, J.C.: Ergodic sets. Bull. Am. Math. Soc. **58**(2), 116–136 (1952)

Chapter 9
Lyapunov Exponent

9.1 Definitions

Consider a discrete dynamic system generated by a diffeomorphism $f : M \to M$ of a compact manifold M. The differential Df is a linear mapping that in local coordinates is a partial derivative matrix:

$$Df(x) = \left(\frac{\partial f_i(x)}{\partial x_j} \right).$$

The differential acts on the tangent bundle

$$Df(x) : TM_x \to TM_{f(x)}$$

and determines a linear extension of the discrete system $x_{n+1} = f(x_n)$ to a tangent bundle, according to the following formula:

$$v_{n+1} = Df(x_n)v_n. \tag{9.1}$$

Every non-degenerate linear mapping $A : R^d \to R^d$ can be represented as

$$Av = |Av|\frac{Av}{|Av|} = r|Ae|\frac{Ae}{|Ae|} = r|Ae|A_s(e),$$

where $r = |v|, |v| \neq 0, e = {}^{v}\!/_{|v|} \in S^{d-1}, S^{d-1}$ is the unit sphere. Thus, a linear mapping $A : v \to Av$ is the product of these two mappings:

$$e \to A_s(e) = \frac{Ae}{|Ae|} \text{ and } r \to r|Ae|,$$

G. Osipenko, *Symbolic Analysis of Dynamical Systems*, Mathematical Marvels: Texts and Monographs in the Spirit of CR Rao, https://doi.org/10.1007/978-981-95-0713-9_9

where the first one acts on the sphere S^{d-1}, and the second one acts on a positive half-line R^+. A projective space P^{d-1} is a set of the one-dimensional subspaces in R^d. The projective space P^{d-1} can be obtained by identifying antipodal points of the sphere S^{d-1}. Symmetry of the mapping $A_s(e)$ with respect to the sign change $A_s(\pm e) = \pm A_s(e)$ allows us to define a mapping $PA(e) = Ae/|Ae|$ onto the projective space P^{d-1} by identifying antipodes points on the sphere.

Let us use the notation e for the points in space P^{d-1}; that is, e denotes a one-dimensional subspace and/or a unit vector in that subspace, without causing confusion.

We get the mapping:

$$Pf(x, e) = \left(f(x), \frac{Df(x)e}{|Df(x)e|} \right)$$

on the projective bundle $P = \left\{ (x, e) : x \in M, e \in P^{d-1}(x) \right\}$, and this mapping generates a discrete dynamical system of the following form:

$$x_{n+1} = f(x_n), \tag{9.2}$$

$$e_{n+1} = \frac{Df(x_n)e_n}{|Df(x_n)e_n|} \tag{9.3}$$

on the projective bundle P.

Positive number $a(x, e) = |Df(x)e|$ is the coefficient of length change on subspace e under the action of differential at point x. The projective bundle P is a compact manifold with metric (see [1]), which can be interpreted as the sum of distance on manifold and angle between subspaces.

Definition 9.1 *Let $\xi = \{(x_k, e_k), k = 0, 1, 2, \cdots\}$ be a semi-trajectory of the system (9.2, 9.3). The upper limit*

$$\lambda(\xi) = \overline{\lim_{n\to\infty}} \frac{1}{n} \ln\left|Df^n(x_0)e_0\right| = \lim_{n\to\infty} \sup_{k>n} \frac{1}{k} \ln\left|Df^k(x_0)e_0\right| \tag{9.4}$$

is called the Lyapunov exponent for the semi-trajectory.

For the value $|Df^n(x_0)e_0|$, the following equalities are true:

$$\left|Df^n(x_0)e_0\right| = \left|Df\left(f^{n-1}(x_0)\right)Df^{n-1}(x_0)e_0\right|$$
$$= \left|Df(x_{n-1})\frac{Df^{n-1}(x_0)e_0}{\left|Df^{n-1}(x_0)e_0\right|}\right| \left|Df^{n-1}(x_0)e_0\right| = \left|Df(x_{n-1})e_{n-1}\right| \left|Df^{n-1}(x_0)e_0\right|.$$

Using this recurrent formula, we obtain the equation

$$\left|Df^n(x_0)e_0\right| = \prod_{k=0}^{n-1} |Df(x_k)e_k|,$$

which allows us to write the Lyapunov exponent in the form:

$$\lambda(\xi) = \overline{\lim_{n\to\infty}} \frac{1}{n} \sum_{k=0}^{n-1} \ln|Df(x_k)e_k|. \tag{9.5}$$

This implies that there is a subsequence $n_m \to \infty$ such that the limit

$$\lambda(\xi) = \lim_{n_m\to\infty} \frac{1}{n_m} \sum_{k=0}^{n_m-1} \ln|Df(x_k)e_k| \tag{9.6}$$

exists and coincides with the Lyapunov exponent $\lambda(\xi)$. The resulting formulas allow us to determine the Lyapunov exponent of a pseudo-trajectory and consider it as an average of the function φ (x,e) = ln |Df(x)e| over pseudo-trajectories.

If $\xi = \{(x_k, e_k)\}$ is a pseudo-trajectory of period p, then all indicated limits exist and are equal to the arithmetic average of ln |Df(x)e| by period:

$$\lambda(\xi) = \lim_{n\to\infty} \frac{1}{n} \sum_{k=0}^{n-1} \ln|Df(x_k)e_k| = \frac{1}{p} \sum_{k=1}^{p} \ln|Df(x_k)e_k| = \frac{1}{p} \sum_{k=1}^{p} \varphi(x_k, e_k). \tag{9.7}$$

Definition 9.2 *The Morse spectrum is an average spectrum of function* $\varphi(x, e) = ln|Df(x)e|$ *over the pseudo-trajectories of system* (9.2, 9.3).

It follows from Proposition 8.1 that the Morse spectrum coincides with the limit set of average values of the function $\varphi(x, e) = ln|Df(x)e|$ over periodic ε-trajectories as $\varepsilon \to 0$:

$$\Sigma = \{\lambda : \textit{there is a sequence of periodic } \varepsilon_n - \textit{trajectories } \xi_n,$$

$$\varepsilon_n \to 0 \textit{ such that } \lambda = \lim_{n\to\infty} \lambda(\xi)\}.$$

As a result of the previous chapter, each component Ω of the chain-recurrent set of mapping *Pf* on the projective bundle generates an interval $[\lambda_{min}(\Omega), \lambda_{max}(\Omega)]$ in the spectrum. These intervals can have a non-trivial intersection for different components of the chain-recurrent set *CR*(*Pf*) [1]. Each component Ω_{ik} of the chain-recurrent set on the projective bundle P is projected onto some component Ω_i of the chain-recurrent set on manifold M. The number of chain-recurrent components on M is not limited. For example, an equation $x' = x^2 sin\left({}^1/_x\right)$ has an infinite number of equilibria. If the mapping f is a shift per unit time along trajectories of this continuous system, the equilibriums of this system coincide with chain-recurrent components of a discrete system generated by this mapping. The number of components Ω_{ik} on projective bundle that are projected onto a chosen component $\Omega_i \subset M$ does not exceed the dimension of the manifold M, see [2].

Let Ω be a component of the chain-recurrent set of Pf, and let $\mathcal{M}(\Omega)$ be the set of invariant probability measures concentrated on Ω. The following theorem is a corollary of Theorem 8.5.

Theorem 9.1 *The Morse spectrum is the union of segments.*

$$\Sigma = \bigcup_k [a_{ik}, b_{ik}],$$

each segment being an averaging set of the function $\varphi(x, e) = ln|Df(x)e|$ *over invariant measures concentrated on a component* Ω_{ik},

$$[a_{ik}, b_{ik}] = \Sigma(\Omega_{ik}) = \left\{ \lambda = \int_M \varphi \, d\mu, \ \mu \in \mathcal{M}(\Omega_{ik}) \right\}.$$

Thus, in order to construct a Morse spectrum with some accuracy, we need to estimate an infinite number of intervals that is generally impossible. However, in our case, these intervals are compressed, as in the example above. This allows us to construct an estimate a finite number of intervals, some of them contain infinitely many spectral intervals. It is clear that the number of approximating intervals can be increased to infinity for $\varepsilon \to 0$. The main tool for calculating the spectrum is a symbolic image of a dynamic system. Theorem 8.3 allows us to estimate the Morse spectrum by using the technique of symbolic image of mapping Pf on the projective bundle.

Let G be the symbolic image of the mapping Pf with respect to a covering $C = \{M(i)\}$ of the projective bundle P, and let $\{H_k\}$ be a complete family of equivalence classes of recurrent vertices of graph G. To calculate the spectrum, it is necessary to convert the function $\varphi(x, e) = ln|Df(x)e|$ to a function $a\big[ij\big]$ on the symbolic image. If $i \to j$ is an edge on G, fixing a point (x, e) in the cell $M(i)$, we set $a\big[ij\big] = ln|Df(x)e|$.

Let H be a class of equivalent recurrent vertices and let

$$\lambda_{min} = min\{\lambda(\psi_j), j = 1, ..., q\},$$

$$\lambda_{max} = max\{\lambda(\psi_j), j = 1, ..., q\}$$

be minimum and maximum average values on cycles of the class H. Since the number of cycles is finite, these values can be calculated in a finite number of steps. To achieve this goal, we can use the linear programming method of [3] or the least characteristic cycle calculation method, see [4]. According to Theorem 8.2, the spectrum of a symbolic image $\Sigma(G)$ consists of the intervals $[\lambda_{min}(H_k), \lambda_{max}(H_k)]$, where $\{H_k\}$ is the complete family of classes of equivalent recurrent vertices in the symbolic image. The following theorem is a corollary of Theorems 8.3.

Theorem 9.2

(1) The Morse spectrum is contained in the extended spectrum of the symbolic image.

$$\Sigma^* = \bigcup_k [\lambda_{min}(H_k) - \eta(d), \lambda_{max}(H_k) + \eta(d)],$$

where $\{H_k\}$ *is the complete family of classes of equivalent recurrent vertices,* d *is the diameter of the covering, and* η *is the continuity modulus of function* $ln|Df(x)e|$.

(2) *If the covering diameter tends to zero, then the extended spectrum* Σ^* *tends to the Morse spectrum in the Hausdorff metric.*

9.2 Exponential Estimates

Suppose we have calculated the extended spectrum Σ^* of the symbolic image. Let us apply the results obtained to evaluate the action of the differential along the trajectories of the system.

Let $C = \{M(i)\}$ be a covering of the projective bundle and let G be a symbolic image of mapping Pf. Recall that there is a multivalued mapping h that maps a point (x, e) to the indices of the cells in which this point lies. If (x, e) is a chain-recurrent point, then $h(x, e)$ are recurrent vertices. The image $h(\Omega)$ of a component of a chain-recurrent set lies in some class equivalent recurrent vertices. Therefore, if Ω is a component of a chain-recurrence set, then there exists a unique component $H(\Omega)$ containing $h(\Omega)$. Consider the ε-trajectory $\chi = \{(x_k, e_k)\}$ of mapping Pf. According to shadowing Theorem 2.1, if $\varepsilon < r$ (r is the lower boundary of the symbolic image), then there is an admissible path $\omega = \{i_k : (x_k, e_k) \in M(i_k)\}$, which is tracked by the pseudo-trajectory χ on the symbolic image.

Theorem 9.3 *(1) Let* $[a, b]$ *be an extended Morse spectrum for the class* H*. Then there exist positive constants* K_* *and* K^* *such that for any finite* ε*-trajectory* $\chi = \big\{(x_0, e_0), (x_1, e_1), ..., \big(x_p, e_p\big)\big\}$, $\varepsilon < r$*, whose admissible path* $\omega = \{i_k : (x_k, e_k) \in M(i_k)\}$ *lies in* H*, the following estimates hold:*

$$K_* exp(pa) \leq \prod_{k=0}^{p-1} |Df(x_k)e_k| \leq K^* exp(pb). \tag{9.8}$$

(2) Let Ω *be a component of the chain-recurrent set for the mapping* Pf *and* $[a, b]$ *be the extended spectrum for the class* $H(\Omega)$*, then*

$$K_* exp(pa)|v| \leq \left|Df^p(x)v\right| \leq K^* exp(pb)|v|, \tag{9.9}$$

for $(x, e) \in \Omega$ *and a vector* $v \in e$.

Proof 1. Let [a, b] be an extended spectrum for the class H. First, we consider the ε-trajectory χ, which is tracked by the cycle $\omega = \{i_1, ...i_p = i_0\}$ of the class H. Average

$$\lambda(\omega) = \frac{1}{p}\sum_{k=0}^{p-1} a\left[i_k i_{k+1}\right]$$

lies in the spectrum of class H. The number $a[ij] = ln\left|Df\left(x_i^*\right)e_i^*\right|$ is determined by the selected point (x_i^*, e_i^*) in the cell $M(i_k)$, and if (x_k, e_k) is another point in $M(i_k)$, then the following inequality holds:

$$|ln|Df(x_k)e_k| - ln|Df\left(x_i^*\right)e_i^*|| \leq \eta(d).$$

Average:

$$a \leq \frac{1}{p}\sum_{k=0}^{p-1} ln|Df(x_k)e_k| \leq b.$$

Therefore, we have an estimate

$$exp\,(pa) \leq \prod_{k=0}^{p-1}|Df(x_k)e_k| \leq exp(pb)$$

for the cycle $\chi = \{(x_0, e_0), (x_1, e_1), ..., (x_p, e_p)\}$. Consider an arbitrary ε-trajectory $\chi = \{(x_0, e_0), (x_1, e_1), ..., (x_p, e_p)\}$, which is tracked by $\omega = \{z_0, z_1, ..., z_p\}$, lying in the class H. We can decompose ω into a sum of simple paths and the remainder as follows. If the path ω passes through the vertex z^* twice, that is, $z^* = z_l = z_{l+s}$, then two paths are formed: $\omega_1 = \{z_0, ..., z_l = z_{l+s} \cdots, z_p\}$ and $\omega_2 = \{z_l \cdots, z_{l+s}\}$. The path ω_2 is the periodic of period s. The path ω is a sum of the paths ω_1 and ω_2. Repeating the decomposition process, we can represent the path ω as a sum of simple periodic paths (cycles) ψ_j from H and a residual path ω_0 with different vertices of the class H. So, we obtain the decomposition:

$$\omega = k_1\psi_1 + k_2\psi_2 + ... + k_q\psi_q + \omega_0,$$

where the integer $k_j \geq 0$ is the number of times the cycle ψ_j repeats in the decomposition of path ω. If $p_1, p_2, \cdots, p_q$ are the periods of cycles and p_0 is the length of the path ω_0, then

$$p = k_1 p_1 + k_2 p_2 + ... + k_q p_q + p_0.$$

Since the vertices of the residual path ω_0 are different, p_0 does not exceed number of vertices in class H.

Let us consider the product

$$\Pi(\omega) = \prod_{k=0}^{p-1} \alpha[z_k], \tag{9.10}$$

where $\alpha[z_k] = |Df(x_k)e_k|$, $(x_k, e_k) \in M(z_k)$. Let us rearrange the product factors according to decomposition:

$$\omega = k_1\psi_1 + k_2\psi_2 + ... + k_q\psi_q + \omega_0.$$

We obtain the representation of product $\Pi(\omega)$ in form:

$$\Pi(\omega) = \prod_{j=1}^{q} \prod_{i=1}^{k_j} \Pi_i(\psi_j) \cdot \Pi(\omega_0).$$

We have estimate

$$exp(p_j a) \leq \Pi_i(\psi_j) \leq exp(p_j b) \tag{9.11}$$

for each factor $\Pi_i(\psi_j)$. Remainder $\Pi(\omega_0)$ has no more t cofactors, and each cofactor admits the estimations

$$K_{min} \leq \alpha[z_k] \leq K_{max}, \ \alpha[z_k] = |Df(x_k)e_k|, \ (x_k, e_k) \in M(z_k)$$

on compact set. Without loss of generality, assume that $K_{min} \leq 1 \leq K_{max}$. Thus, the remainder $\Pi(\omega_0)$ has estimations

$$(K_{min})^t \leq \Pi(\omega_0) \leq (K_{max})^t. \tag{9.12}$$

From (9.11) and (9.12) follows estimate

$$K_* exp(pa) \leq \Pi(\omega) \leq K^* exp(pb), \tag{9.13}$$

where

$$K_* = \begin{cases} (K_{min})^t \exp(-at), npu\, a > 0, \\ (K_{min})^t, npu\, a \leq 0, \end{cases} \tag{9.14}$$

$$K^* = \begin{cases} (K_{max})^t \exp(-bt), npu\, b < 0 \\ (K_{max})^t, npu\, b \geq 0. \end{cases}$$

The required inequalities (9.8) follow from what was said before.

2. Proof of the estimate (9.9). Since trajectory is an ε-trajectory for any $\varepsilon > 0$, estimate (9.8) is valid for every trajectory χ lying in Ω. If $(x_0, e_0) = (x, e) \in \Omega$ and $(x_1, e_1) = Pf(x, e)$, the basis vector has the form:

$$e_1 = \frac{Df(x_0)e_0}{|Df(x_0)e_0|}.$$

Therefore, we get

$$|Df(x_1)e_1|\left|Df(x_0)e_0\right| = \left|Df(f(x))\frac{Df(x)e}{|Df(x)e|}\right|\left|Df(x)e\right| = \left|Df^2(x)e\right|.$$

In the same way, we can obtain equality

$$\prod_{k=0}^{p-1}|Df(x_k)e_k| = \left|Df^p(x)e\right|,$$

where $e = v/|v|$. In this case, the inequalities (9.8) take form:

$$K_* exp(pa)|v| \leq \left|Df^p(x)v\right| \leq K^* exp(pb)|v|.$$

Theorem is proven.

9.3 Hyperbolicity

Let $E_1, E_2, ...E_l$ be subspaces of a linear space E. The direct sum of these subspaces $E = E_1 \oplus ... \oplus E_l$ means that $E = E_1 + ... + E_l$ and $E_i \cap E_j = \{0\}$ for any $i \neq j$.

Definition 9.3 *A smooth dynamic system $x_{n+1} = f(x_n)$ is called hyperbolic on an invariant set $\Lambda \subset M$, if there exist invariant subbundles E^s and E^u of the tangent space $TM|_\Lambda$, the constants $K > 0$ and $\alpha > 0$ such that*

$$TM|_\Lambda = E^s \oplus E^u,$$

$$\left|Df^n(x)v\right| \leq K|v|exp(-\alpha n), x \in \Lambda, v \in E^s(x), n > 0,$$

$$\left|Df^{-n}(x)v\right| \leq K|v|exp(-\alpha n), x \in \Lambda, v \in E^u(x), n > 0,$$

where the invariance of the subbundle E^* means the equality:

$$Df(x)E^*(x) = E^*(f(x)).$$

The number α is called the hyperbolic exponent.

Theorem 9.4 [5, 6]. *Let $x_{n+1} = f(x_n)$ be a smooth system on a compact manifold M, Ω is a component of chain-recurrent set on M, and $\pi : TM \to M$ is the projection from a tangent bundle to the manifold. Then*

(1) *The chain-recurrent set of the restriction $Pf(x,e) = \left(f(x), \frac{Df(x)e}{|Df(x)e|}\right)$ on $P|_\Omega$ has l components $\Omega_1, \ldots, \Omega_l$, $1 \le l \le dimE$.*
(2) *Every set Ω_k defines a (continuous, constant dimension) subbundle E_k over Ω of the form:*

$$E_k = \left\{ v \in \pi^{-1}(\Omega) : v \neq 0 \Rightarrow \frac{v}{|v|} = e \in \Omega_k \right\}.$$

(3) *Decomposition of the tangent bundle takes the form:*

$$TM|_\Omega = E_1 \oplus \ldots \oplus E_l.$$

(4) *Every chain-recurrent component Ω^* in the bundle $P|_\Omega$ is projected onto the chain-recurrent component Ω and has the properties described in the item (2).*

According to Theorem 9.4, components $\Omega_1, \ldots, \Omega_l$ uniquely determine the decomposition $TM|_\Omega = E_1 \oplus \ldots \oplus E_l$. Since each chain-recurrent component is an invariant set, the decomposition is also invariant. The decomposition mentioned in Theorem 9.4 is called the finer Morse decomposition of tangent bundle.

Let $C(M) = \{m(j)\}$ be a covering of a manifold M and let $C(P) = \{M(z)\}$ be a conformal covering of projective bundle P, that is, each cell's projection is a cell. We denote the symbolic images of the mappings f and Pf by $G(f)$ and $G(Pf)$, respectively. The equality $\pi M(z) = m(j)$ leads to a natural mapping $\pi^*(z) = j$ from $G(Pf)$ to $G(f)$, which is a mapping of directed graphs. Let $\Omega \subset M$ and $\Omega_k \subset P$ be the components described in Theorem 9.4 and $H_k = H(\Omega_k)$.

Then the projection $\pi^* H_k = H(\Omega)$. The following theorem is a corollary of Theorem 9.3.

Theorem 9.5 Test for hyperbolicity. *If 0 is not in the extended spectrum of the classes $\{H_k\}$, then the diffeomorphism f is hyperbolic on Ω.*

Proof First, consider the following simple observation. Suppose that a linear map A satisfies the inequality $c|v| \le |Av|$ and there is an inverse map A^{-1}, then $|A^{-1}v| \le {}^1/_c|v|$. Indeed, if we substitute $w = A^{-1}v$ in the inequality $c|w| \le |Aw|$, then we obtain the desired inequality. Therefore, the inequality

$$K_* exp(na)|v| \leq |Df^n(x)v|, \ n > 0$$

follows from the inequality

$$|Df^{-n}(x)v| \leq \frac{1}{K_*} exp(-na)|v|, n > 0.$$

Since 0 is outside the extended spectrum, each class H_k has a negative extended spectrum $[a_k, b_k]$, $b_k < 0$ or a positive extended spectrum $[a_k, b_k]$, $a_k > 0$. Consider all the classes H_k with a positive extended spectrum $[a_k, b_k]$, $a_k > 0$. Denote $a = min\{a_k > 0\} > 0$. According to Theorem 9.3, every class H_k generates an invariant E_k for which the inequality

$$K_* exp(na_k)|v| \leq \left|Df^n(x)v\right| \leq K^* exp(pb_k)|v|$$

satisfies, where $n > 0, v \in E_k$. Let the unstable subbundle E^u be the sum of subspaces E_k with positive extended spectrum. Then the inequality

$$\left|Df^{-n}(x)v\right| \leq Kexp(-na)|v|$$

holds for $v \in E^u, n > 0$ and some $K \geq 1$. Similarly, the classes H_k with negative extended spectrum $[a_k, b_k]$, $b_k < 0$ generate the subbundles E_k and the stable subbundle E^s, which is the sum of subspaces E_k with a negative extended spectrum. Inequality

$$\left|Df^n(x)v\right| \leq Kexp(nb)|v|$$

holds for $v \in E^s$, $n > 0$ and $b = max\{b_k < 0\} < 0$. From Theorem 9.4 it follows

$$TM|_\Omega = E^u \oplus E^s$$

that the theorem is true.

Section 10.5 gives an example of calculating the extended Morse spectrum using a computer and checking hyperbolicity for a non-trivial invariant set.

9.4 Ω-Stability

Definition 9.4 *A point x is called non-wandering if for any neighborhood U of the point x there exists $N > 0$ such that $f^N(U) \cap U \neq \emptyset$. The set of the non-wandering points is denoted by Ω.*

From this definition, it follows that non-wandering points are chain-recurrent: $\Omega \subset CR$. The following example shows that this inclusion is strict.

Example 9.1 Consider a system generated by a differential equation $\varphi' = cos^2\varphi$ on the unit circle. A shift per unit time defines a discrete dynamical system. The system has two fixed points $\varphi = \frac{\pi}{2}$ and $\varphi = \frac{3\pi}{2}$, forming a set Ω of non-wandering points. The derivative φ' is positive for $\varphi \neq \frac{\pi}{2}, \frac{3\pi}{2}$, so $\varphi(t)$ increases and the trajectories go counterclockwise. The chain-recurrent set consists of a single component that matches the circle $S \supset \Omega$.

Let us introduce a distance into the space of continuous mappings. Let $f : M \to M$ and $g : M \to M$ be continuous mappings, and let

$$d_0(f, g) = \sup_M \rho(f(x), g(x))$$

be the C^0-distance. If the mappings f and g are smooth, then

$$d_1(f, g) = d_0(f, g) + \sup_M |Df(x) - Dg(x)|$$

is the C^1-distance. Suppose that the chain-recurrent set is hyperbolic. In this case, the Anosov closure lemma [7] states that $\Omega = CR$. Periodic trajectories are dense in this set, and a following theorem on the structural stability of hyperbolic sets holds.

Theorem 9.6 [7, 8] *Let a compact invariant set Λ be hyperbolic for a diffeomorphism f. Then, for any neighborhood V of the set Λ and any $\delta > 0$ there exists $\varepsilon > 0$ such that, if $d_1(f, g) < \varepsilon$, then there is a hyperbolic invariant set $\Lambda(g) \subset V$ for g and a homeomorphism $h : \Lambda(g) \to \Lambda$ satisfying the conditions.*

$$h \circ g|_{\Lambda(g)} = f|_{\Lambda} \circ h,$$

and $d_0(Id, h) + d_0\left(Id, h^{-1}\right) < \delta$, where Id is the identity mapping and the homeomorphism h is unique for sufficiently small δ.

The theorem is purely local in character; it says nothing about dynamics outside the neighborhood of the invariant set Λ. This theorem can be applied both to the chain-recurrent set *CR* as a whole and to its isolated components.

A question arises: if the chain-recurrent set *CR* is hyperbolic for f and g is C^1-close to f, what can be said about the chain-recurrent set $CR(g)$ and the dynamics on it? Before answering this question, we have to note that a similar question for the non-wandering set Ω leads to a non-trivial answer: the non-wandering set can increase sharply with C^1-perturbation of the diffeomorphism f, even if the non-wandering set Ω is hyperbolic. This phenomenon is called the Ω-explosion and an example of the described phenomenon is given in the book [8].

Definition 9.5 *A diffeomorphism f is Ω-stable if there is its C^1-neighborhood U such that for each diffeomorphism $g \in U$ there exists a homeomorphism $h : \Omega(g) \to \Omega(f)$ such that $f \circ h = h \circ g$.*

For Ω-stability, in addition to the hyperbolicity of the non-wandering set, an additional condition is required—the absence of cycles (see [8]). In this case, the non-wandering set coincides with the chain-recurrent set. Let us return to the properties of the chain-recurrent set *CR*. In book [5] it was shown that the mapping $f \to CR(f)$ is upper semicontinuous in C^0-topology. More precisely, if x_n are chain-recurrent points of the maps f_n, $x_n \to x_0$, and $f_n \to f_0$, in the C^0-topology, then x_0 is a chain-recurrent point of f_0. Thus, the chain-recurrent set does not allow explosions with C^0-perturbations of the system. However, the next example shows that the chain-recurrent set can decrease sharply with a perturbation of the system.

Example 9.2 Consider the differential equation $\varphi' = cos\varphi$ on the unit circle. The equation has two equilibria: $N(\varphi = \frac{\pi}{2})$ is stable and $S(\varphi = \frac{3\pi}{2})$ is unstable. The remaining trajectories go from S to N. Let $h(\varphi)$ be a positive function for all φ except $\varphi = 0$, $h(0) = 0$. Equation of the form $\varphi' = (h(\varphi) + \varepsilon)cos\varphi$, $\varepsilon > 0$ has the same equilibrium states as the original system. However, when $\varepsilon = 0$ the equation receives an additional equilibrium state $\varphi = 0$, which is < <passing> >through. If f is a shift by unit of time along the trajectories of the equation $\varphi' = (h(\varphi) + \varepsilon)cos\varphi$, $\varepsilon \geq 0$, then the discrete system $\varphi_{n+1} = f_\varepsilon(\varphi_n)$ for $\varepsilon = 0$ has a chain-recurrent set consisting of three fixed points $\varphi = 0, \pi/2, 3\pi/2$, and the equilibrium state $\varphi = 0$ disappears as $\varepsilon > 0$. Thus, the chain-recurrent set sharply decreases when system is perturbed.

It follows from Theorem 9.6 that if a chain-recurrent set is hyperbolic, then the chain-recurrent set and the dynamics on it are stable under C^1-perturbation. As already noted, if a chain-recurrent set is hyperbolic, then the equality $\Omega = CR$ holds. Consequently, a system is Ω-stable if its chain-recurrence set is hyperbolic.

Theorem 9.7 *The following statements are equivalent:*

(1) *the diffeomorphism f is hyperbolic on the chain-recurrent set CR;*
(2) *the Morse spectrum contains no zero;*
(3) *there exists $d_0 > 0$ such that the extended spectrum of the symbolic image for a covering of a projective bundle with maximal diameter $d < d_0$ does not contain zero.*

Proof The implication (2) $\Rightarrow$ (1) follows from Theorems 9.2 and 9.5. Let us prove the implication (1) $\Rightarrow$ (2). Let f be hyperbolic on the chain-recurrent set and let α be an exponent of the hyperbolicity. Let us show that the Morse spectrum Σ does not contain an interval $[-\beta, \beta]$, where $0 < \beta < \alpha$. Consider periodic ε-trajectories $\chi = \{(x_k, e_k\}$ on the projective bundle with $\varepsilon \to 0$. We can consider χ in the chain-recurrent set $CR(P)$ on the projective bundle. Let P^s and P^u be the projections onto projective bundle of stable E^s and unstable E^u subbundles. According to the papers [5, 9], the unstable subbundle P^u is an attractor, and the stable subbundle P^s is its dual repeller. Then the chain-recurrent set $CR(P)$ must lie in $P^u \cup P^s$. Since P^u and P^s are invariant closed sets, we can assume that the pseudo-trajectories χ lie either in P^u, or in P^s. Let $\chi = \{(x_1, e_1), ...(x_N, e_N)\}$ be an ε-trajectory of length N lying in P^s. We fix a positive integer $q > 0$, define $N = mq + r$, $0 \leq r < q$ and split the pseudo-trajectory χ into m pieces of length q and the remainder of length r. The compactness of P implies the estimate:

$$\left| ln \prod_{i=0}^{q} |Df(x_i)e_i| - \ln\left|Df^q(x_1)e_1\right| \right| \leq \delta(q, \varepsilon),$$

where q is fixed and $\delta(q, \varepsilon)$ tends to zero at $\varepsilon \to 0$ uniformly with respect to (x_i, e_i). By the definition of hyperbolicity, we have

$$\left|Df^q(x)e\right| \leq Kexp(-\alpha q),$$

where $e \in E^s$. The exponent of pseudo-trajectory $\chi = \{(x_k, e_k)\}$ has the form:

$$\begin{aligned}
\lambda(\chi) =& \frac{1}{N} ln\left(\prod_{k=1}^{N} |Df(x_k)e_k|\right) \\
=& \frac{1}{N} ln\left(\prod_{p=0}^{m-1}\left(\prod_{k=pq+1}^{(p+1)q} |Df(x_k)e_k|\right)\right) \prod_{k=mp+1}^{N} |Df(x_k)e_k| \\
=& \frac{1}{N}\left(\sum_{p=0}^{m-1} ln \prod_{k=mp+1}^{(p+1)q} |Df(x_k)e_k| + \sum_{k=mq+1}^{N} ln|Df(x_k)e_k|\right).
\end{aligned}$$

Subtracting and adding the expression $\frac{1}{N}\sum\limits_{p=0}^{m-1} ln\left|Df^q\left(x_{pq+1}\right)e_{pq+1}\right|$, we get

$$\begin{aligned}
&\frac{1}{N}\sum_{p=0}^{m-1}\left(ln \prod_{i=1}^{q} \left|Df\left(x_{(p-1)q+i}\right)e_{(p-1)q+i}\right| - ln\left|Df^q\left(x_{pq+1}\right)e_{pq+1}\right|\right) \\
&\quad + \frac{1}{N}\sum_{k=mq+1}^{N} ln|Df(x_k)e_k| + \frac{1}{N}\sum_{p=0}^{m-1} ln\left|Df^q\left(x_{pq+1}\right)e_{pq+1}\right| \leq \\
&\quad \frac{m}{N}\delta(q, \varepsilon) + \frac{rC}{N} + \frac{m}{N}(lnK - \alpha q),
\end{aligned}$$

where $|Df| \leq C$. We have

$$\begin{aligned}
\lambda(\chi) &\leq -\alpha + \frac{N - mq}{N}\alpha + \frac{lnK}{q} + \delta(q, \varepsilon) + \frac{rC}{N} \\
&= -\alpha + \frac{lnK}{q} + \delta(q, \varepsilon) + \frac{r(C + \alpha)}{N} \leq -\alpha + \frac{lnK}{q} + \delta(q, \varepsilon) + \frac{C + \alpha}{m}.
\end{aligned}$$

Let us choose a q such that $\frac{lnK}{q} < \frac{\alpha - \beta}{6}$. Having q given we choose $\varepsilon > 0$ so that

$\delta(q, \varepsilon) < \frac{\alpha-\beta}{6}$. Finally, let m be such a large number that $\frac{C+\alpha}{m} < \frac{\alpha-\beta}{6}$. Thus, we get

$$\lambda(\chi) \leq -\alpha + \frac{\alpha - \beta}{2} < -\beta.$$

Similarly, we can show that if $\chi \in P^u$, then

$$\lambda(\chi) \geq \alpha - \frac{\alpha - \beta}{2} > \beta.$$

Thus, we have the equivalence (1) $\Leftrightarrow$ (2) holds.

From the second assertion of Theorem 8.3, it follows that the Hausdorff distance between Morse spectrum and the extended spectrum of the symbolic image tends to zero as the covering diameter d converges to zero. Taking into account the separation of these spectra from zero, we obtain the equivalence (2) $\Leftrightarrow$ (3). The proof is complete.

9.5 Homoclinic Tangency

Let f be a diffeomorphism on a plane with a hyperbolic fixed point O. Let us denote by W^s and W^u the stable and unstable manifolds of the point O. Suppose that there is a point $Q \neq O$ at which these manifolds are tangent to each another. This means that the point Q is homoclinic, and the tangent spaces $TW^s(Q)$ and $TW^u(Q)$ coincide. We will call this case a homoclinic tangency. An example of a mapping with such a property is given by

$$f(x, y) = (bx + y + ax(1 - x), y + ax(1 - x)),$$

where a = 1.511717565, b = 0.93. The dynamical system $(x_{n+1}, y_{n+1}) = f(x_n, y_n)$ has a fixed point (0, 0). The stable and unstable manifolds for this point are shown in Fig. 9.1. Since $detDf(x, y) = b < 1$, we are in the conditions of the Newhouse's phenomenon (see [10]), the Newhouse's phenomenon occurs. In any neighborhood of a dissipating diffeomorphism exhibiting a non-degenerate homoclinic tangency, there exists a residual set of diffeomorphisms that admit an infinite number of periodic sinks. The paper [11] gives an example of a homoclinic tangency, the dynamics of which have the following properties.

There is a sequence of periodic trajectories converging to a homoclinic trajectory. These periodic trajectories are hyperbolic, and their Lyapunov characteristic exponents are uniformly separated from zero. However, the angle between the stable and unstable subspaces is not separated from zero, which is natural for a homoclinic tangency. There are known examples of diffeomorphisms that have stable periodic

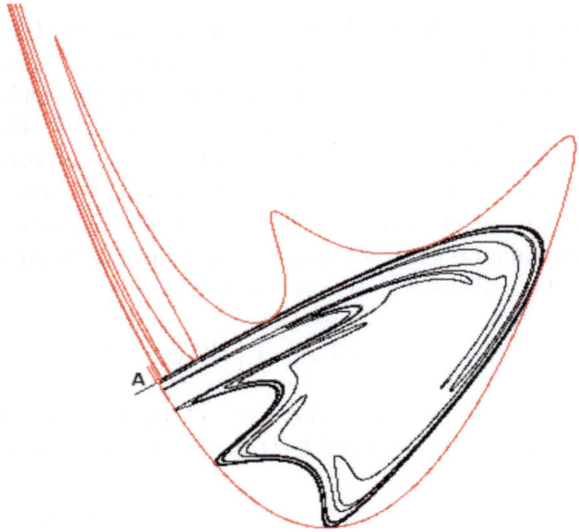

Fig. 9.1 Homoclinic tangency

trajectories in any neighborhood of the homoclinic trajectory, and the Lyapunov exponents for these periodic trajectories are separated from 0 (see [12]).

Let us show that in the case of homoclinic tangency, a chain-recurrent set in the projective bundle contains the entire layer P^1 over the hyperbolic point O and, as a result, the Morse spectrum contains the segment $[\lambda_s, \lambda_u]$, where λ_s and λ_u are the stable and unstable characteristic exponents at the point O.

Theorem 9.8 *Let f be a diffeomorphism of a plane with a hyperbolic fixed point O and its stable and unstable manifolds tangent at the homoclinic point. Then*

(1) *the chain-recurrent set of the mapping Pf on the projective bundle contains the product $\Omega \times P^1$, where Ω is the closure of the homoclinic trajectory, P^1 is an one-dimensional projective space;*
(2) *the Morse spectrum of the mapping f contains the segment $[\lambda_s, \lambda_u]$, where λ_s and λ_u are the stable and unstable Lyapunov exponents at the point O.*

Proof Consider a chain-recurrent set CR of the mapping

$$Pf(x, e) = \left(f(x), \frac{Df(x)e}{|Df(x)e|}\right)$$

on the projective bundle $P = R^2 \times P^1$. The distance on P^1 is the angle between the subspaces, which can be identified with the unit vectors P^1 defined up to a sign. The proof of the theorem consists of several propositions.

Proposition 9.1 *The chain-recurrent set $CR(Pf)$ on the projective bundle contains a one-dimensional projective space $P^1(O)$, that is, for any $\varepsilon > 0$ and any subspace $e \in P^1(O)$ there is a periodic $\varepsilon > 0$-trajectory of the mapping Pf passing through e.*

Proof Over the fixed point O, the mapping $Pf(O)$ acts on $P^1(O)$, where it has two fixed subspaces e^s and e^u, which are tangent spaces to the stable and unstable manifolds. In addition, e^s is the source (repeller) and e^u is the sink (attractor) dual to e^s. All other trajectories start at e^s and end at e^u. Consider a subspace e different from e^s and e^u. The trajectory $Tr(O, e)$ of the subspace e starts at e^s and ends at e^u. The homoclinic trajectory $Tr(Q)$ of the point Q starts and ends at a hyperbolic point O. Consider the trajectory $Tr(Q, e^*)$ of the subspace $e^* = TW^s(Q) = TW^u(Q)$ (over the point Q) under the action of the projective mapping Pf. From the invariance of stable and unstable subspaces follows that

$$Tr(Q, e^*) = \{TW^s(x) = TW^u(x), x \in Tr(Q)\}.$$

This trajectory starts at $O \times e^u$ and ends at $O \times e^s$. Thus, we have constructed two trajectories $Tr(O, e)$ and $Tr(Q, e^*)$ that form a cycle of the form:

$$(O, e^s) \to (O, e) \to (O, e^u) \to (Q, e^*) \to (O, e^s).$$

Now we have to prove that for any $\varepsilon > 0$ there exists a periodic ε-trajectory passing through e. Let the distance ρ on the projective bundle $P = R^2 \times P^1$ be defined as maximum distance on plane and angle between the subspaces. Let U be the ε-neighborhood of the hyperbolic point O. Since the trajectory $Tr(Q)$ starts and ends at O, there exists positive integer k such that $f^{-k}(Q) \in U, f^0(Q) = Q$, and $f^k(Q) \in U$. Number k may be large enough. Let us choose this number so that the angles $\angle(Df^{-k}e, e^s) < \varepsilon/2$ and $\angle(Df^k e, e^u) < \varepsilon/2$. Since trajectory $Tr(O, e)$ of the subspace e starts at e^s and ends at в e^u, there exists integer n such that the angles $\angle(Df^{-n}e, e^s) < \varepsilon/2$ and $\angle(Df^n e, e^u) < \varepsilon/2$. Hence

$$\rho((f^{-k}(Q), Df^{-k}e^*), (O, Df^n e)) < \varepsilon,$$

$$\rho((f^k(Q), Df^k e^*), (O, Df^{-n}e)) < \varepsilon$$

that allows us to construct a periodic ε-trajectories of the form

$$\chi^* = \{(O, Df^i e), -n \le i \le n; (f^j(Q), Df^j e^*), -k \le j \le k\}$$

on the projective bundle. These pseudo-trajectories pass through subspace e, have period 2(k + n), the numbers k and n are determined by the value ε. Since e is an arbitrary subspace in $O \times P^1$ and is different from e^s and e^u, entire circle $O \times P^1$ consists of chain-recurrent points. Proof is over.

Remark. From proof it follows that the trajectory $Tr(Q, e^*), e^* = TW^s(Q) = TW^u(Q)$ and the circle $O \times P^1$ lie in the chain-recurrent set of mapping Pf.

Let f be above-described diffeomorphism with hyperbolic fixed point and homoclinic trajectory. Denote λ_s and λ_u Lyapunov exponents of the hyperbolic point O, $\lambda_s < 0 < \lambda_u$.

Preposition 9.2 *If the stable and unstable manifolds of a hyperbolic point are tangent at a homoclinic point, then the Morse spectrum includes the segment* $[\lambda_s, \lambda_u]$.

Proof While proving the previous proposition, we constructed two trajectories $Tr(O, e)$ and $Tr(Q, e^*)$ forming a cycle:

$$\left(O, e^s\right) \to (O, e) \to \left(O, e^u\right) \to \left(Q, e^*\right) \to \left(O, e^s\right).$$

Using this cycle, we construct a periodic pseudo-trajectory. The stable and unstable subspaces e^s and e^u of the hyperbolic point are fixed points of the mapping Pf. The points $Tr(O, e^s) = (O, e^s)$ and $Tr(O, e^u) = (O, e^u)$ are the trajectories of $Pf.Pf$.

This allows us to include these points in the constructed cycle. Thus, we obtain a periodic ε-trajectory of the form:

$$\chi = \{\left(O, Df^i e\right), -n \le i \le n; \left(O, Df^p e^u\right), 0 \le p \le N;$$

$$\left(f^j(Q), Df^j e^*\right), -k \le j \le k; \left(O, Df^q e^s\right), 0 \le q \le K\}.$$

This pseudo-trajectory passes through the subspace e and has a period $2(k+n)+N+K$, where the numbers k and n are fixed and determined by the value ε, N and K are arbitrarily large positive integers. Let us calculate the Lyapunov exponent of the constructed pseudo-trajectory:

$$\lambda(\chi) = \frac{1}{2(k+n)+N+K} \sum_{z=0}^{2(k+n)+N+K-1} \ln |Df(x_z)e_z). \tag{9.15}$$

The sum consists of three parts: the first part Σ_0 contains the items generated by the pseudo-trajectory χ^* and has 2(k + n) terms, the second part Σ_s is generated by the action of the differential on the stable subspace e^s and has K terms, and the last part Σ_u is generated by the action of the differential on unstable subspace e^u and has N terms. The numbers k and n are fixed, the number N and K will be chosen large enough so that $\lambda(\chi)$ will approximate a number in the segment $[\lambda_s, \lambda_u]$. The first part is limited $\Sigma_0 = C$, the second part is $\Sigma_s = K\lambda_s$, and the third part is $\Sigma_u = N\lambda_u$.

Thus, we have

$$\begin{aligned}(\chi) =& \frac{1}{2(k+n)+N+K}(C + K\lambda_s + N\lambda_u) = \\ =& \frac{C}{2(k+n)+N+K} + \left(\frac{N+K}{2(k+n)+N+K}\right)\left(\frac{1}{N+K}\right)(K\lambda_s + N\lambda_u).\end{aligned}$$

For $N + K \to \infty$

$$\frac{C}{2(k+n)+N+K} \to 0 \text{ and } \frac{N+K}{2(k+n)+N+K} \to 1.$$

Let us take $0 < \alpha < 1$ and construct a sequence of positive numbers N and K such that $N + K \to \infty$ and $\frac{K}{N+K} \to \alpha$. If α is a rational number, then we choose N and K as a solution to the equation:

$$\frac{K}{N+K} = \alpha,$$

that is, we take $\frac{N}{K} = \frac{1-\alpha}{\alpha}$. There are infinitely many such numbers and we can choose them such that $N + K \to \infty$. If α is an irrational number, then it is sufficient to take $\frac{N}{K} \to \frac{1-\alpha}{\alpha}$ and $N + K \to \infty$. For such a choice of N and K, we have

$$\lambda(\chi) \to \alpha\lambda_s + (1-\alpha)\lambda_u,$$

that is, the Morse spectrum contains the number $\alpha\lambda_s + (1-\alpha)\lambda_u$. If α is an arbitrary number between 0 and 1, the Morse spectrum contains any number from the segment $[\lambda_s, \lambda_u]$. The proof is complete.

Consider the projective space $P^1(Q)$ over the homoclinic point Q. There exists a subspace $e^* = TW^s(Q) = TW^u(Q)$, which has the following properties. If v is a vector in e^*, then

$$\left|Df^n v\right| \to 0 \, npu \, n \to +\infty$$

and

$$\left|Df^n v\right| \to 0 \, npu \, n \to -\infty.$$

If $e \in P^1(Q)$ is a subspace different from e^* and w is a vector in e, then

$$\left|Df^n w\right| \to \infty \, npu \, n \to +\infty$$

and

$$\left|Df^n w\right| \to \infty \, npu \, n \to -\infty.$$

Preposition 9.3 The chain-recurrent set $CR(Pf)$ on the projective bundle contains a one-dimensional projective space $P^1(Q)$.

Proof We need to prove that for any $\varepsilon > 0$ and any subspaces $e \in P^1(Q)$ there exists a periodic ε-trajectory of the mapping Pf passing through e. Since O is a hyperbolic point of f, the mapping Pf has two fixed points $e^s = TW^s(O)$ and $e^u = TW^u(O)$.

Moreover, all other trajectories of the mapping $Pf(O)$ start at e^s and end at e^u. Let $e \in P^1(Q)$ be a subspace other than $e^* = TW^u(Q) = TW^s(Q)$.

Then the iterations of $Pf^n e$ are such that

$$Pf^n e \to e^u \; npu \; n \to +\infty$$

and

$$Pf^n e \to e^s \; npu \; n \to -\infty.$$

Consequently, there is a trajectory of the form:

$$e^s \to e \to e^u.$$

In addition, there is also the trajectory of homoclinic tangency:

$$e^u \to e^* \to e^s.$$

Combining these two trajectories, we obtain the cycle

$$e^s \to e \to e^u \to e^* \to e^s,$$

which twice passes over the closure of the homoclinic trajectory. Using the constructed cycle and acting similar to Proposition 9.1, we construct a periodic pseudo-trajectory passing through e. The proof is complete.

The validity of Theorems 9.8 follows from Propositions 9.1–9.3.

References

1. Colonius, F., Kliemann, W.: The Dynamics of Control. Birkhauser (2000)
2. Osipenko, G.: Dynamical systems, graphs, and algorithms. Lectures Notes in Mathematics, 1889, Springer, Berlin (2007)
3. Osipenko, G.S., Romanovsky, J.V., Ampilova, N.B., Petrenko, E.I.: Computation of Morse spectrum. J. Math. Sci. **120**(2), 1155–1166 (2004)
4. Georgiadis, L., Goldberg, A.V., Tarjan, R.E., Werneck, R.F.: An experimental study of minimum mean cycle algorithms. In: Proceedings 6th International Workshop on Algorithm Engineering and Experiments, ALENEX 2009, pp. 1–13 (2009)
5. Bronshtein, I.U.: Non-autonomous Dynamical System. Kishiniv, Shtinitsa (1984) (in Russian)
6. Selgrade, J.: Isolated invariant sets for flows on vector bundles. Trans. Amer. Math. Soc. **203**, 359–390 (1975)
7. Katok, A., Hasselblatt, B.: Introduction to the Modern Theory of Dynamical Systems. Cambridge University Press (1995)
8. Nitecki, Z.: An Introduction into the Orbit Structure of Diffeomorphisms. The MIT press Cambridge (1971)
9. Franke, J., Selgrade, J.: Hyperbolicity and chain recurrence. J. Differ. Equ. **26**, 27–36 (1977)

10. Newhouse, S.: Diffeomorphisms with infinitely many sink points. Topology **12**, 9–18 (1974)
11. Cao, Y., Luzzatto, S., Rios, I.: Some non-hyperbolic systems with strictly non-zero Lyapunov exponents for all invariant measures: horseshoes with internal tangencies. Discret. Contin. Dynam. Syst. **15**(1), 61–71 (2006)
12. Vasil'eva, E.V.: Smooth diffeomorphisms of the plane with stable periodic points in a neighborhood of a homoclinic point. Differ. Equ. **48**(3), 309–317 (2012)

Chapter 10
Numerical Experiments

10.1 Computer-Oriented Technology

The theoretical results of the previous chapters provide the rationale for using a symbolic image to numerically study the global dynamics of a system. Chapters 2 and 3 show that paths on symbolic images approximate trajectories of a dynamical system under Tikhonov topology. Based on this, computer-oriented algorithms have been designed to study the global structure of the trajectories of dynamic systems. Section 3.5 describes a method for localizing a chain-recurrent set and gives an example of this construction for a hyperbolic recurrent set of the Ikeda mapping. Chapter 4 describes results that allow to calculate attractors and their basins. It has been shown that the construction of filtration is based on canonical sorting of symbolic image. In Chaps. 5 and 6, it has been shown that the space of flows on symbolic image approximates the set of invariant measures of dynamical system in the weak topology. This forms the basis for numerical construction of invariant measures and determining their properties. Section 5.7 describes the balance method, allowing to construct an invariant measure with certain extremal properties. Chapter 7 provides formulas for estimation of metric entropy through the entropy of flows on symbolic images, including a method for computing the maximum entropy measure. Chapter 8 is devoted to the spectrum of averaging a function over pseudo-trajectories of dynamical system, showing that this spectrum is approximated by spectrum of averaging over the periodic paths on symbolic image. Calculating the spectrum of symbolic image reduces to the problem of averaging on extreme cycles, which is solved within the framework of graph theory. In Chap. 9, the developed theory is applied to calculating Morse spectra, showing, in particular, that a dynamical system is hyperbolic when its Morse spectrum does not contain zero. This allows numerical verification of hyperbolicity of a dynamical system.

G. Osipenko, *Symbolic Analysis of Dynamical Systems*, Mathematical Marvels: Texts and Monographs in the Spirit of CR Rao, https://doi.org/10.1007/978-981-95-0713-9_10

The General Scheme of Using Symbolic Image

Suppose we want to solve a problem for a dynamical system. Let us transform the system into a symbolic image by choosing a finite covering of the phase space. This transforms the original problem into a graph theory problem, which then leads to a specific task for the symbolic image. We use methods and algorithms from graph theory to solve the problem for the graph. Next, we need to interpret the solution, converting the information about the graph back into information about our dynamical system. As a rule, this information is an approximation of the original solution. The accuracy of this information depends on the diameter of the covering. To achieve the desired accuracy, we construct a sequence of coverings and symbolic images by successive subdivisions. It is important to establish criteria to stop further subdivisions. Any exploration into symbolic images forms part of the basic computational structure. Here, we describe how this structure is implemented and discuss aspects that need to be considered for efficient implementation. The framework proposed here was implemented and tested in the AnT project, see [1].

Symbolic image can be seen as an iterative process that can be repeated in order to increase the discretization level of the phase space at each step. At each stage, there are three main steps that need to be repeated:

(1) Subdivision of selected cells of the phase space into smaller cells.
(2) Construction of a symbolic image for the current level of discretization. Reformulating the dynamical system problem as a graph problem.
(3) Application of graph theory methods to solve the graph problem and interpreting the results for the dynamical system. This allows us select parts of space for further subdivision and more precise analysis.

Thus, the circuit operates according to the following cycle:

As shown in Fig. 10.1, these steps will be repeated until a termination criterion is reached. The end depends on desired accuracy and available computing power. The termination criteria can be determined based on the task. For example, if the problem is to determine whether a system is hyperbolic, the process will stop when zero is not in the extended spectrum of the symbolic image. The user can terminate the process at their discretion.

In the theory of symbolic image, there are no restrictions on the geometry of the investigated domain and the cells, except that they have to be compact. In general, a domain M can be any compact of the phase space and must be provided by the user. Here, we study a rectangular region M. Then, we subdivide the area M into grid boxes $M(i)$. The construction of a symbolic image from numerical calculations is only an estimation of the "real" symbolic image. Algorithms used for an efficient calculation of symbolic images are discussed in the papers [1–3]. There are several issues that need to be addressed when constructing a cell image. Cells are typically selected as rectangular parallelopipeds. To construct edge $i \to j$, we need to check that

$$f(M(i)) \bigcap M(j) \neq \emptyset.$$

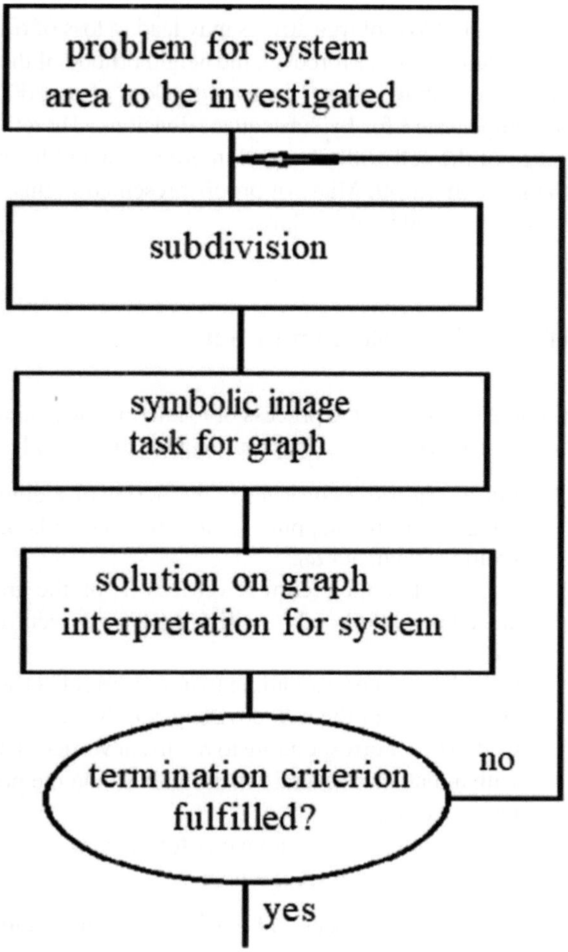

Fig. 10.1 Flowchart of the described method by symbolic analysis

The image of each cell $f(M(i))$ can be calculated approximately. For our implementation, we use a simple method to approximate a cell image. Each cell $M(i)$ is approximated by a finite set of so-called scan points. The number N of scan points and their positions within the cells are parameters that must be specified by the user, but there is no rule for how scan points should be placed within a cell. Our experience shows that a good strategy is to distribute the scan points uniformly within a cell. Number scan point N is determined by the balance between counting time and the accuracy. Usually, for smooth systems, this is $10 \leq N \leq 100$ sufficient. We believe that an arrow $i \to j$ exists if there is at least one scan point $x_k \in M(i)$ such that $f(x_k) \in M(j)$. The dissertation [3] discusses five methods of constructing a cell image, but the problem solved by the user may determine how the symbolic image will be constructed. For example, if we are localizing a chain-recurrent set CR, then the extra arrows can add recurrent vertices and increase neighborhood

$U \supset CR$, but loss of true arrows may lead to loss of recurrent vertices and part of the chain-recurrent set. Moreover, the neighborhood of the desired set can be reduced by successive subdivisions. However, the loss of part of the chain-recurrent set cannot be compensated for by subsequent divisions. Therefore, it is better to construct an image of the cell with a margin in order to avoid losing arrows when localizing the chain-recurrent set. Many of proofs presented in this book are constructive and can be implemented in algorithms.

10.2 Chain-Recurrent Set

Section 3.5 justifies the process of localizing the chain-recurrent components. Based on the results obtained, we propose a localization algorithm.

(1) Let us choose a covering $C = \{M(i)\}$ of a compact set M. Find a symbolic image G of the mapping f. Note that the cells of the initial covering can have arbitrary diameter d_0.
(2) Let us define the recurrent vertices $\{i\}$ on the graph G. Using them, we find a closed neighborhood $P = \left\{\bigcup M(i) \ : \ i \text{ is recurrent}\right\}$ of the chain-recurrent set CR.
(3) Cells $\{M(j)\}$ corresponding to nonrecurrent vertices $\{j\}$ do not participate in the construction of the new covering and the new symbolic image.
(4) Cells $M(i)$, corresponding to recurrent vertices $\{i\}$, are divided into smaller cells with a diameter several times smaller than the previous one. Thus, we define a new covering.
(5) Construct a symbolic image G for the new covering.
(6) Let's move to the second item.

Repeating the process of successive refinement of coverings, according to Theorem 3.9, we obtain a sequence of nested neighborhoods $P_0, P_1, P_2, \cdots$ of the chain-recurrent set CR and a sequence of covering diameters $d_0, d_1, d_2, \cdots$ of the cells corresponding to recurrent vertices. By construction, $d_{k+1} \leq \lambda d_k$, $0 < \lambda < 1$ and, therefore, $d_k \to 0$. Thus, the described algorithm gives a monotonic decreasing sequence of neighborhoods that converges to a chain-recurrent set.

Implementation of the Algorithm

Let $G(V, E)$ be a directed graph with a set V of vertices, and a set E of edges. It has been shown that the problem of localizing the chain-recurrent set of a dynamic system reduces to studying the corresponding symbolic image and identifying recurrent vertices. As noted, the classes of equivalent recurrent vertices in graph theory are known as strong connectivity components (see Sect. 11.4 or references [4, 5]). The algorithms have a linear time complexity, and the number of operations estimated as O(n + m), where n is the number of vertices and m is the number of edges. Moreover, these algorithms are implemented as codes and are available on the Internet. The method for localizing chain-recurrent set was applied to the Julia mapping.

Example 10.1 In 1918, a 25-year-old Frenchman Gaston Julia healed his wounds at the hospital. Tired of idleness, he began to study the behavior of points of a sequence $z_{n+1} = z_n^2 + C$ on a complex plane. The Julia set is a set of points in complex plane that do not go to infinity during the described iterations. It turned out to be dependent on a constant number C and has a complex fractal structure. Fractals exhibit self-similarity at increasingly smaller scales. Now Julia sets are well studied and beautiful images are created that are used for design. Our goal is to find the chain-recurrent set of the Julia mapping, which is part of the Julia set but may not be identical to it. The Julia mapping generates a system of difference equations on a plane in the form of

$$x_{n+1} = x_n^2 - y_n^2 + a,$$

$$y_{n+1} = 2xy + b.$$

Note: This mapping is not a homeomorphism and it is an example of using a symbolic image for semi-dynamical systems. Here, we give numerical results for a = b = −0.5, $M = [-2, 2]\times[-2, 2]$.

The Julia set is a closed two-dimensional region Ω with a non-trivial fractal boundary for these parameters, see Fig. 10.2. To find the chain-recurrent set, we used the algorithm described earlier. We chose an initial covering of the compact set M with squares of side h = 0.25. Five subdivisions were created and a neighborhood of the chain-recurrent set was found. This invariant set has two components: one Ω_1 coincides with the boundary of the Julia set and the other Ω_2 is located near the origin. Using the standard iterative process, we can show that Ω_1 is a repeller, and Ω_2 is an attractor. Therefore, any trajectory outside $\Omega_1 \cup \Omega_2$ starting at the boundary Ω_1 and ends at the attractor Ω_2.

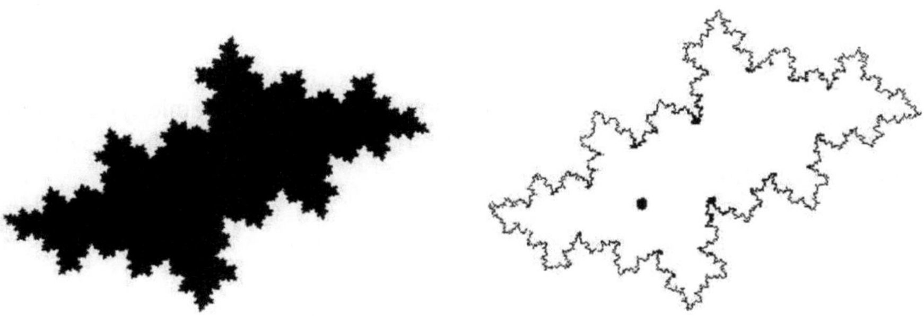

Fig. 10.2 The Julia set, a = −0.5, b = −0.5, and the chain-recurrent set

10.3 Attractors

It will be shown here that the results from Chap. 4 can be used as computer-oriented algorithms. Consider a discrete dynamical system $x_{n+1} = f(x_n)$ on a compact set M. Let $C = \{M(i)\}$ be a closed covering of this set and G be a symbolic image with respect to C. According to Proposition 2.2, the vertices of the symbolic image can be renumbered so that the transition matrix takes the form:

$$\Pi = \begin{pmatrix} \Pi_1 \cdots\cdots\cdots\cdots \\ \ddots \\ 0 \;\; \Pi_k \cdots\cdots \\ \ddots \;\; \ddots \\ 0 \;\; 0 \;\; \Pi_s \end{pmatrix},$$

where each diagonal block Π_k either corresponds to one equivalence class of recurrent vertices H_k, or corresponds to nonrecurrent vertex consisting of one zero. Section 11.5 describes a method for such renumbering. Attractors in a symbolic image are found as follows. Let us consider a transition matrix as described above, where each class of equivalent recurrent vertices H_k produces a diagonal block Π_k. With this numbering, every class H_k generates a set of vertices

$$L = \{i \geq min(j \in H_k)\}.$$

The structure of the admissible transition matrix implies that it is possible to enter L, but not to leave L. This means that L is an attractor in the symbolic image. The closed set

$$A = \{\cup M(i), i \in L\}$$

is an absorbing neighborhood of some attractor of the original dynamic system. From Propositions 2.2 and 4.2, it follows that a matrix of admissible transitions with this numbering of vertices is represented as

$$\Pi = \begin{pmatrix} \Pi^* \cdots\cdots\cdots\cdots \\ \ddots \\ 0 \;\; 0 \cdots\cdots \\ \ddots \;\; \ddots \\ 0 \;\; 0 \;\; \Pi \end{pmatrix},$$

where block Π corresponds to attractor L. Zeros located diagonally between blocks Π^* and Π correspond nonrecurrent vertices. Block Π^* corresponds to repeller L^*, dual to L. Diagonal zeros between Π^* and Π correspond to attraction domain

$$D(L) = \{i : max(j \in L^*) < i < min(k \in L)\}$$

in symbolic image. Trajectories of every point in set

$$W = \{\cup M(i), i \in D(L)\}$$

end in absorbing set A. Thus, A is an estimate (upper) of attractor and W is a lower estimate of the attraction domain. Theorems 4.1 and 4.2 form the basis of the following algorithm.

Algorithm for constructing an attractor, its domain of attraction, and repeller for a dynamical system.

(1) Let C be the covering of a compact set M by cells of maximum diameter d. The symbolic image G is constructed.
(2) Let us choose an attractor L on the graph G, its attraction domain $D(L)$, and a repeller L^*.
(3) Obtain sets

$$A = \{\cup M(i), i \in L\}, W = \{\cup M(j), j \in D(L)\},$$

$$R = \{\cup M(k), k \in L^*\}.$$

Let $d_1 = max\{diamM(i), diamM(k) : i \in L, k \in L^*\}$.

(4) Subdivide the cells corresponding to the attractor L and the repeller L^*. The remaining cells do not change. Create a new covering.
(5) Construct a symbolic image of the new covering.
(6) Construct a new attractor NL and a repeller NL^* as the maximal invariant sets in L and L^*, respectively. Put $L := NL$, $L^* := NL^*$, $D = Ver(NG)\backslash L^*$.
(7) Go to step 3.

By repeating this process, we obtain a sequence of sets

$$A_1, A_2, \cdots ; W_1, W_2, \cdots ; R_1, R_2, \cdots ,$$

and sequence of numbers $d_1, d_2, \cdots$.

Theorem 10.1 [6]. The described algorithm determines

(1) The sequences of sets that are nested into each other

$$A_1 \supset A_2 \supset \cdots ,$$

$$W_1 \subset W_2 \subset \cdots ,$$

$$R_1 \supset R_2 \supset \cdots .$$

(2) If $d_s \to 0$ as $s \to \infty$, then

$$\lim_{s\to\infty} A_s = \bigcap_s A_s = \Lambda \quad \text{is an attractor,}$$

$$\lim_{s\to\infty} W_s = \bigcup_s W_s = W^s(\Lambda) \quad \text{is its domain of attraction,}$$

$$\lim_{s\to\infty} R_s = \bigcap_s R_s = \Lambda^* \quad \text{is the repeller dual to } \Lambda$$

(3) Each attractor Λ can be constructed using the described algorithm.

Generally speaking, the number of possible attractors on the symbolic image increases with decreasing covering diameter d. Moreover, their structure can be quite complex, and their number can be unlimited for $d \to 0$. Filtration describes the structure of these attractors. Section 4.6 justifies an algorithm for constructing a refined sequence of filtrations.

Example 10.2 *Attractors of the Duffing Equation.*

Let us consider an example of the numerical construction of an attractor and its attraction domain for the Duffing equation. The differential equation of the Duffing oscillator is given by

$$m\ddot{x} = ax + bx^3 + c\dot{x} + dcos\omega t,$$

where x and m are the coordinate of the particle and its mass, respectively, and $F = dcos\omega t$ is the action of an external periodic force. The equation was studied by German Engineer Georg Duffing 1918. Let us consider the equation for $a = 1$, $b = -1$, $c = -0.15$ and without external force (d = 0). We introduce the notation $\dot{x} = y$ and get a system of differential equations:

$$\begin{aligned} \dot{x} &= y, \\ \dot{y} &= x - x^3 + cy. \end{aligned} \tag{10.1}$$

We can construct a homeomorphism $f : R^2 \to R^2$ generated by this system by numerically constructing a mapping $f(x, y)$, which represents a shift per unit time along the trajectories of the system. The dynamics is studied in the domain $M = [-2, 2]\times[-1.5, 1.5]$. The system has three equilibrium states (0,0), (−1,0), (1,0). The first point is hyperbolic, while the others are attractors. The maximal attractor in the given domain is the closure of unstable manifold W^u of the point

(0,0). This manifold $W^u(0,0)$ ends at the points A (−1,0) and B (1,0), and there are no other attractors in this domain. For attractor A, we can estimate its fundamental neighborhood and area of attraction. The initial covering C consisted of squares with side 0.5.

To obtain a better localization of the attractor and to estimate its fundamental neighborhood, seven subdivisions were required (each square was split into four sub-squares). As a result, see Fig. 10.3, 2653 cells were needed to obtain an estimate of the attractor and 113,424 cells were needed for estimating the area of attraction. The maximal attractor is the closure of unstable manifold $W^u(0,0)$ of the hyperbolic equilibrium (0, 0). An estimate of the maximal attractor in this area was also obtained, which is also shown in Fig. 10.4. Here we have filtration of attractors of the form:

$$\emptyset \subset A \subset (A \cup B) \subset \overline{W^u(0,0)}\} \subset R^2.$$

Other examples of using symbolic images for numerical construction of attractors and their attraction domains can be found in papers [1, 7].

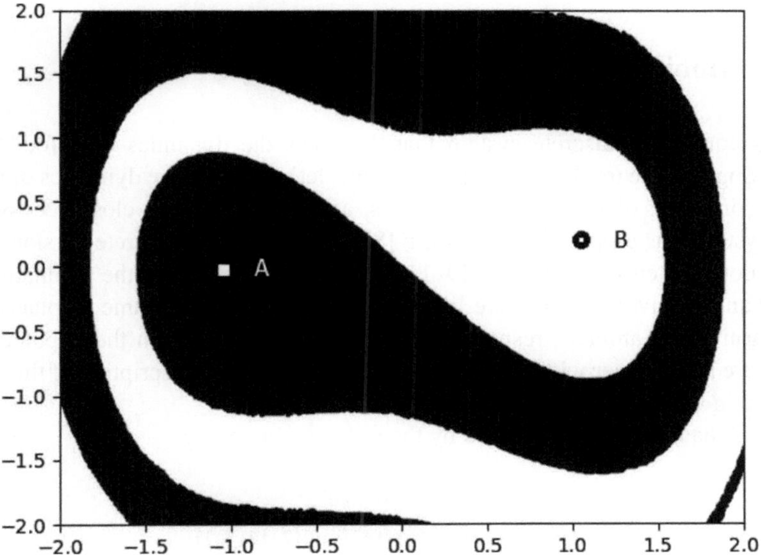

Fig. 10.3 Estimation of the attractor A and its area of attraction

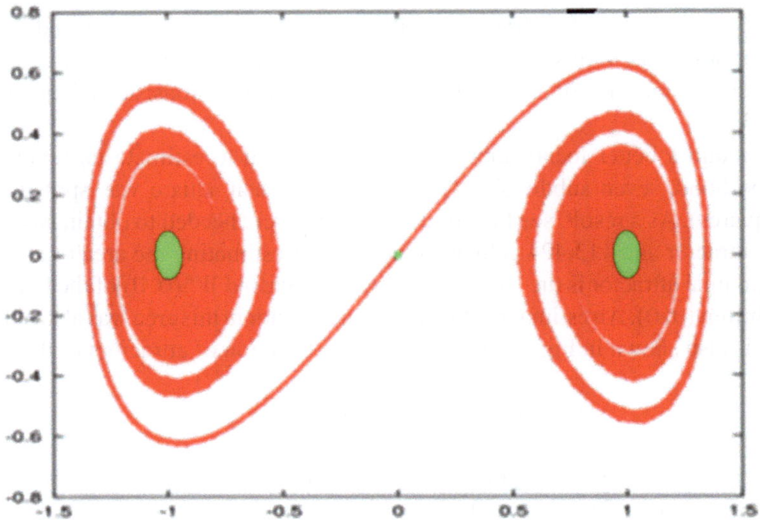

Fig. 10.4 Estimation of the maximal attractor

10.4 Mobius Strip as an Attractor

Let us consider a discrete system that describes the dynamics of a food chain consisting of three trophic levels. For example, let's consider the dynamics of a food chain consisting of vegetation, herbivores, and carnivores in a closed ecosystem. This system was proposed by Linstrom [8] as a generalized discrete version of the continuous systems described in [9–11]. It can be assumed that the populations of plants and carnivore x and z are linearly dependent on the volume of plant foods and carnivorous animals, respectively. The dependence of y on the population of herbivore is nonlinear with a positive derivative. A detailed description of the model is given in article [12].

The dynamic system is defined by the following mapping:

$$x_{n+1} = \frac{M_0 x_n e^{-y_n}}{1 + x_n max(e^{-y_n}, g(z_n)g(y_n))},$$

$$y_{n+1} = M_1 x_n y_n e^{-z_n} g(y_n) g(M_2 y_n z_n),$$

$$z_{n+1} = M_2 y_n z_n,$$

where

$$g(s) = \begin{cases} \frac{1-e^{-s}}{s}, & if \ \ s \neq 0, \\ 1, & if \ \ s = 0. \end{cases}$$

By design, dynamics is considered for positive values $x, y,$ and z. This system has been studied in [1, 2, 6, 7], where the bifurcations of the topological structure and the transition to chaotic behavior when parameters change are considered. The system has four equilibrium states: E_0 is the origin, E_1 lies on the x-axis, E_2 lies in the positive quadrant of the $xy-$plane, and E_3 lies in the positive quadrant of $xyz-$space (for details see [12]). Here we will look at the application of the methods described above. Let us show the appearance of an attractor of this system in the shape of a Möbius strip. Let us fix $M_1 = 1$, $M_2 = 4$ and change M_0 from 2 to 3.4. If M_0 varies from 2 to 2.9, then the equilibrium state E_3 is stable and represents a single attractor in $x > 0, y > 0, z > 0$, and the coordinates of E_3 are continuously changing.

At $M_0 \approx 2.93$, the equilibrium state E_3 loses stability and a Neimark–Sacker bifurcation occurs. As a result, a stable invariant curve C appears, which is a boundary of two-dimensional unstable manifold $W^u(E_3)$ of the equilibrium state E_3. The use of symbolic image allows us to localize E_3 and the curve C filled with chain-recurrent orbits. We found an extended spectrum of average of the logarithm of the differential determinant $ln|det(Df)|$, which lies in negative segment $[-0.58055, -0.186926]$ at $3 \leq M_0 \leq 4$. This means that the volume (under influence of system) exponentially tends to zero; therefore, all attractors have zero volume.

At $M_0 \approx 3.366$, the invariant curve C loses stability and a stable invariant curve C^* appears near it, which is twice as long. This is called a bifurcation, similar to the doubling of period—"Feigenbaum bifurcation." The curve C has two-dimensional stable and unstable manifolds. The boundary of the unstable manifold $W^u(C)$ becomes an attractor. Let us show that at $M_0 = 3.3701$ the unstable manifold $W^u(C)$ is a Mobius band, which is an attractor. See Figs. 10.5 and 10.6.

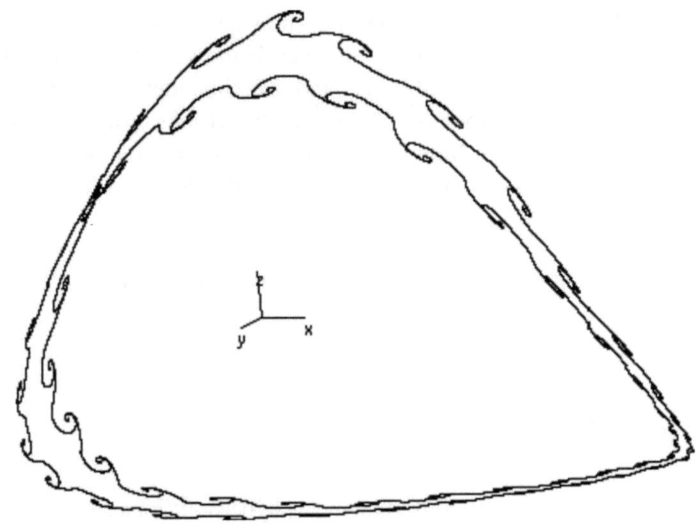

Fig. 10.5 Mobius strip as an attractor, $M_0 = 3.3701$

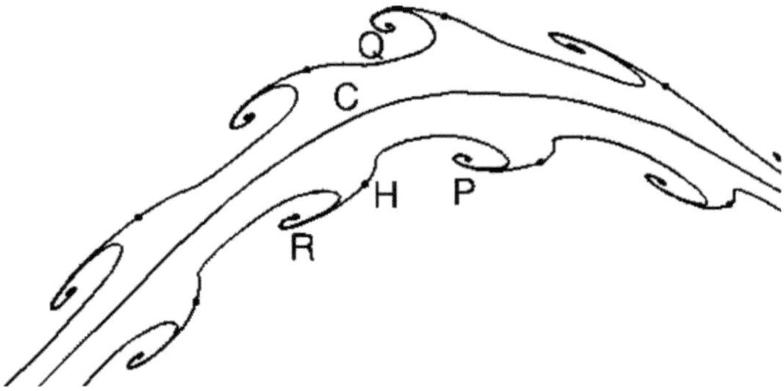

Fig. 10.6 Detail of the Mobius strip; P, Q, R are points of a 55-periodic stable trajectory; H is the point of a 55-periodic hyperbolic trajectory; C is an unstable invariant curve that is the central axis of the Mobius strip

On the bounder $C^* = \partial W^u(C)$ of the unstable manifold there is a stable 55-periodic trajectory $Tr(P)$ of the point $P(1.5193, 0.1402, 0.8471)$, which can be localized using a symbolic image. Points $R(1.6320, 0.1058, 0.7788)$ and $Q(1.4568, 0.1577, 0.7184)$ also lie on this trajectory, as shown in Fig. 10.6.

The hyperbolic 55-periodic trajectory $Tr(H)$ of the point $H(1.5823, 0.1184, 0.8152)$ lies on the boundary $C^* = \partial W^u(C)$. Calculating the eigenvalues of the differential Df^{55} at point P shows that it is a focus on $W^u(C)$. We will now monitor the direction of focus rotation along the trajectory $Tr(P)$. If we follow along the trajectory $Tr(P)$ from P to R, etc., the direction of rotation remains the same, as seen in Fig. 10.6. However, if we compare the directions of rotation at points Q and P (Fig. 10.6), which lie on opposite sides of the center line C, these rotations will be opposite. This can only happen if the unstable manifold $W^u(C)$ is unoriented, that is, $W^u(C)$ is a Mobius band. The bounder of $W^u(C)$ forms the closure of the unstable manifold $\overline{W^u(Tr(H))}$ of hyperbolic trajectory $Tr(P)$, see Figs. 10.5 and 10.6. Thus, the Mobius band $W^u(C)$ is an attractor containing the closure $\overline{W^u(Tr(H))}$, which contains a stable trajectory $Tr(P)$. As a result, we have filtration of attractors

$$\emptyset \subset Tr(P) \subset \overline{W^u(Tr(H))} \subset W^u(C) \subset \{x > 0, y > 0, z > 0\}.$$

10.5 Hyperbolicity Test

According to Theorem 9.7, a diffeomorphism f is hyperbolic on a chain-recurrent set if and only if the Morse spectrum contains no zero or the extended spectrum of the symbolic image of the differential Df does not contain zero. The extended spectrum consists of the intervals

$$[a_k, b_k] = [\lambda_{min}(H_k) - \eta(d), \lambda_{max}(H_k) + \eta(d)],$$

where $\{H_k\}$ is a complete family of classes of equivalent recurrent vertices, d is the diameter of the covering, and η is the continuity modulus of the function $ln|Df(x)e|$. By Theorem 8.2,

$$\lambda_{min}(H) = min\{\lambda(\varphi_j), j = 1, ..., q\},$$

$$\lambda_{max}(H) = max\{\lambda(\varphi_j), j = 1, ..., q\}$$

are the minimum and the maximum of the averages of the function $ln|Df(x)e|$ over cycles $\{\varphi_j\}$ of class H. The described periodic paths are called extremal cycles on the framed graph. The equipment of an edge $i \to j$ is determined as $a[ij] = ln|Df(x_i)e_i|$, where (x_i, e_i) lies in the cell $M(i)$. The problem of computing extreme cycles on a framed graph is a popular topic in graph theory, see [13–18]. Section 11.6 describes the solution algorithm. The article [19] provides a method for numerically verifying hyperbolicity.

Let us consider the modified Ikeda mapping:

$$\begin{pmatrix} x \\ y \end{pmatrix} \to \begin{pmatrix} d + a\,(x \cos \tau(x, y) - y \sin \tau(x, y)) \\ b\,(x \cos \tau(x, y) + y \sin \tau(x, y)) \end{pmatrix}, \tag{10.2}$$

$$\tau(x, y) = 0.4 - \frac{6}{1 + x^2 + y^2}, \tag{10.3}$$

where $d > 0, -1 < a < 1, 0 < b < 1$. The mapping changes orientation when $d = 2, a = -0.9, b = 0.9$ and has a global attractor in the domain $M = [-10, 10] \times [-10, 10]$ (part of the attractor, see Fig. 10.7).

There is only one hyperbolic fixed point $H(1.3815, -2.4746)$. Moreover, the mapping has a unique 2-periodic trajectory $P_2 = \{(0.2338, -0.7031), (1.9995, 0.6681)\}$, whose stable and unstable manifolds intersect, forming a homoclinic loop. A homoclinic cycle generates an infinite number of periodic trajectories in the attractor. Calculations indicate that the

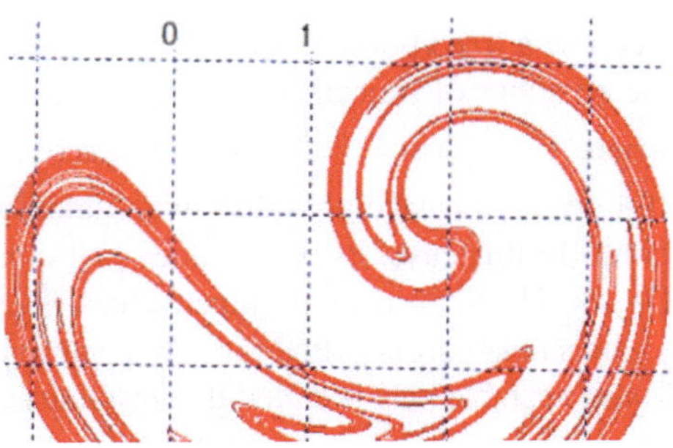

Fig. 10.7 Part of attractor for the modified Ikeda mapping

attractor is not hyperbolic. We want to find a non-trivial hyperbolic invariant subset within the attractor. To do this, we consider part of this attractor, specifically the maximum chain-recurrent subset Ω_0 in the domain $M = [-1.1, 3.5]\times[-1.5, 1.8]$, see Fig. 10.7. Applying the technique to localize a chain-recurrent set yields the following results. The initial covering of the domain M has cells of size 0.46×0.33. After several divisions (with cell size of 0.0071×0.0051), the set Ω_0 splits into two subsets Ω_1 and Ω_2 corresponding to the two strongly connected components in the symbolic image. The invariant subset Ω_1 turned out to be 6-periodic trajectory. For better localization, several divisions were performed until the cell size becomes 0.00045×0.00032.

Construction of the equipped symbolic image was done until the cell angle in projective space became 0.0003. With this data, the distance between Morse spectra of the original system and symbolic image does not exceed 0.01. As a result of calculations on projective bundle, two components are obtained, and Morse spectrum over set Ω_1 is contained in the union of intervals $[0.522, 0.523]$ and $[-0.734, -0.733]$. This means extended spectrum lies in segments $[0.512, 0.533]$ and $[-0.744, -0.723]$, which contain Morse spectrum over invariant set Ω_1. It follows that Ω_1 is 6-periodic hyperbolic trajectory. These calculations were verified by directly calculating Lyapunov exponents of 6-periodic trajectory Ω_1.

The invariant set Ω_2 turned out to be a non-trivial Cantor set. Division is carried out to a cell size 0.00089×0.00058. The construction of the equipped symbolic image was carried out until cell size in the projective space (angle) becomes 0.00038. With such data, the Morse spectrum of the original system differs from the spectrum of the symbolic image by 0.01. As a result, we obtain two intervals $[0.632, 0.793]$ and $[-1.004, -0.843]$, which are extended intervals of equipped symbolic image. The Morse spectra of the set Ω_2 lie in the constructed intervals. These intervals do not contain zero, so set Ω_2 is hyperbolic and system dynamics near set Ω_2 is preserved under C^1-perturbations. For details, see paper [19]. The method of checking the structural stability based on a symbolic image of the dynamic system is described in paper [8].

10.6 Calculation of Homoclinic Trajectories

Let p be a fixed point of a diffeomorphism $f : R^n \to R^n$, that is, $f(p) = p$. To study the behavior of trajectories near p, let us move on to the linear mapping $x \to Ax$, where $A = Df(p) = \left(\frac{\partial f_i}{\partial x_j}(p)\right)$ is the Jacobi matrix of the map f, calculated at point p. If A does not have eigenvalues with modulus is 1, then p is called a hyperbolic fixed point. In this case, its eigenvalues $\{\lambda_i\}$ consist of two parts: the stable eigenvalues $\{|\lambda_i| < 1\}$ and the unstable eigenvalues $\{|\lambda_j| > 1\}$. The eigenspace E^s corresponding to the stable part is called the stable subspace and the eigenspace E^u corresponding to the unstable part is called the unstable subspace. According to the definition,

$$E^s + E^u = R^n, E^s \cap E^u = 0.$$

The Perron theorem describes the structure of local invariant manifolds near a hyperbolic point.

Theorem 10.2 [20]. *Let p be a fixed hyperbolic point of a diffeomorphism f. Then there exist locally stable and locally unstable invariant manifolds*

$$W^s_{loc}(p) = \{x \in U : f^k(x) \to p \ as \ k \to +\infty\}$$

$$W^u_{loc}(p) = \{x \in U : f^k(x) \to p \ as \ k \to -\infty\}$$

in some neighborhood U of point p, with tangent spaces and subspaces E^s and E^u at point p.

The local manifold W^s_{loc} (W^u_{loc}) is uniquely defined in the sense that, if they are constructed for different neighborhoods V_1, V_2, they coincide at the intersection $V_1 \cap V_2$. We can assume that the neighborhood U is a ball with small enough radius, and the local manifolds form disks with dimensions k and $n-k$. Global stable and unstable manifolds for a fixed point p are defined as follows:

$$W^s(p) = \{x \in R^n : f^m(x) \to p \ \ m \to +\infty\} = \left\{\bigcup_m f^m(W^s_{loc}(p)),\ m \in \mathbb{Z}^-\right\},$$

$$W^u(p) = \{x \in R^n : f^m(x) \to p \ \ k \to -\infty\} = \left\{\bigcup_m f^m(W^s_{loc}(p)),\ m \in \mathbb{Z}^+\right\},$$

where $\mathbb{Z}^- = \{0, -1, -2, \cdots\}$, $\mathbb{Z}^+ = \{0, 1, 2, \cdots\}$. In other words, a global stable manifold is the image of a local stable manifold after negative iterations of the mapping f, and a global unstable manifold is the image of a local unstable manifold after positive iterations. The constructed manifolds correspond to one-to-one images of the Euclidean spaces R^k and R^{n-k}.

The invariant manifolds $W^s(p)$ and $W^u(p)$ determine not only the local dynamics near a point p, but also significantly influence the global trajectory structure, especially when stable and unstable manifolds intersect. The points where stable and unstable manifolds intersect $q \in W^s(p) \cap W^u(p)$, $q \neq p$ are called the homoclinic points. Since these manifolds are invariant, the homoclinic points form an invariant set. We can talk about homoclinic trajectories that pass through these points. We say that manifolds W_1 and W_2 intersect transversally at point $q \in W_1 \cap W_2$ if the sum of their tangent spaces $TW_1(q) + TW_2(q)$ is R^n. Next, let's look at the hyperbolic case in R^2. Then stable and unstable manifolds are curves called separatrices. Each separatrix has two branches, dividing a small neighborhood of a hyperbolic point into four sectors.

At the end of the nineteenth century, H. Poincaré studied planetary motion (see paper [21, 22]) and showed that the system has trajectory that tends to a periodic trajectory as time tends to both plus and minus infinity. He called such trajectories double asymptotic. Now they are called homoclinic trajectories. Smile [23] showed that there is an invariant set Ω near a transversal homoclinic trajectory, which can be encoded by all possible sequences of two symbols. This means that chaos occurs on the invariant set Ω. Therefore, the existence of a transversal homoclinic point guarantees chaos. Our goal is to develop a computer-based method for finding a homoclinic point with transverse intersection. Such method was developed in the book [6].

First of all, we note that the unstable manifold of the inverse mapping f^{-1} is a stable manifold of direct mapping f. The construction of a stable manifold resembles the construction of the unstable one. The difference is that we use the direct mapping f to construct an unstable manifold, and we use inverse mapping f^{-1} to create a stable one. Therefore, we will focus on constructing an unstable manifold. Consider a hyperbolic fixed point O of a diffeomorphism $f : R^2 \to R^2$. The Jacobi matrix $A = \left(\frac{\partial f_i}{\partial x_j}(O)\right)$ at O has eigenvalues λ_1, λ_2 and eigenvectors v_1 and v_2, respectively. Since O is a hyperbolic point, the inequality $|\lambda_1| < 1 < |\lambda_2|$ is valid. The stable space $E^s(O)$ is spanned by v_1 and unstable space $E^u(O)$ is spanned by v_2. Let us take a sufficiently small segment L passing through O in the unstable subspace. The unstable separatrix $W^u(O)$ is a smooth image of R, and the point O is the image of the origin. In this case, the segment $L \subset E^u(O)$ touches the separatrix $W^u(O)$ at the point O.

Theorem 10.3 [6]. *Let O be the hyperbolic fixed point of the diffeomorphism f and L be a sufficiently small interval described above. Then the iterations $f^m(L)$ converge to the compact part of unstable manifold $W^u(O)$ in the C^1-topology as $m \to \infty$. In this case, the iterations $f^m(L)$ converge pointwise to the global unstable manifold $W^u(O)$.*

Let us consider an algorithm for the approximate construction of an unstable separatrix using iteration of segment L. In computer processing, it is natural to represent curves as polylines (broken lines) with vertices. In this case, iteration of a curve to be considered as an iteration of vertices of the polyline. The first difficulty lies is increasing the length of the polyline segment during iterations. Therefore, such a broken line cannot approximate a separatrix accurately. To overcome this difficulty, we introduce additional broken vertices as follows. Let us fix a sufficiently small number $h > 0$. Let us construct a broken line such that the length of each polyline segment does not exceed h. The parameter h is selected by the user to achieve the required precision. Let the approximating polyline have a link [AB] and a distance $\rho(f(A), f(B)) > h$ (Fig. 10.8). In this case, let us introduce a point C on the segment [AB] and assume that the approximated broken polyline has two links [AC] and [CB] instead of link [AB]. If the distances $\rho(f(A), f(C)) < h$ and $\rho(f(C), f(B)) < h$, then the division process stops. Otherwise, it is necessary to perform a new division of each link if the image length is greater than h. Thus, a new approximating broken line with a link length less than h is constructed. We get an algorithm for computing

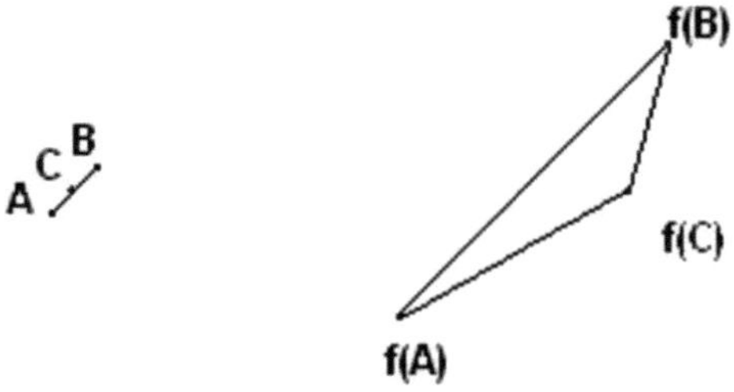

Fig. 10.8 Selecting a polyline link

approximations of unstable manifolds and these approximations are as accurate as computer time allows.

By Theorem 10.3, a segment L touches the unstable manifold. However, for practical computations, the initial segment L can be chosen to be a segment that intersects the stable manifold transversally. The parameter h can be chosen not just from the perspective of accuracy. For a better graphical representation, choose the parameter h so that the resulting polylines look smooth on the screen. It is enough that the h parameter is less than 1 pixel. To avoid unnecessary calculations, exclude the position of the polylines outside the selected neighborhood. The curves in their figures are constructed in this way.

A stable separatrix is created in a similar manner by using iterations of an inverse mapping f^{-1} and a sufficiently small segment L^* passing through the fixed point O and intersecting the unstable manifold in transverse manner. To create a homoclinic point, it two approximations $W_m = f^m(L)$ and $W_m^* = f^{-m}(L^*)$ are needed, and for a good approximation, m must be large enough and segments L and L^* must be small. Next, find the intersection points of W_m and W_m^*, then estimate the angle between them. To find coordinates of the homoclinic point we proceed as follows: find the intersection points W_m and W_m^* different from O, which approximate the homoclinic point. Choose one of these, say H. Thus, we obtain two broken curves W_m[OH] and W_m^*[OH] approximating stable and unstable separatrices. All other parts of the broken lines can be discarded. The length of the unstable curve $f(W_m)$[OH] increases with positive iterations and the length of the stable curve $f^{-1}(W_m^*)$[OH] increases with negative iterations. By repeating this procedure, we get a sequence of points approximating a homoclinic point. For each iteration, we can estimate the angle formed by stable and unstable curves.

As an example, we will use the described method to construct the separatrices of the hyperbolic point of the system generated by the mapping

$$\begin{aligned} x_1 &= x + y + ax(1 - x) \\ y_1 &= y + ax(1 - x) \end{aligned} \tag{10.4}$$

where $a = 1.35$. The fixed point A (0,0) is a hyperbolic point. To construct an inverse mapping for mapping (10.4), this system must be solved for x and y. This gives

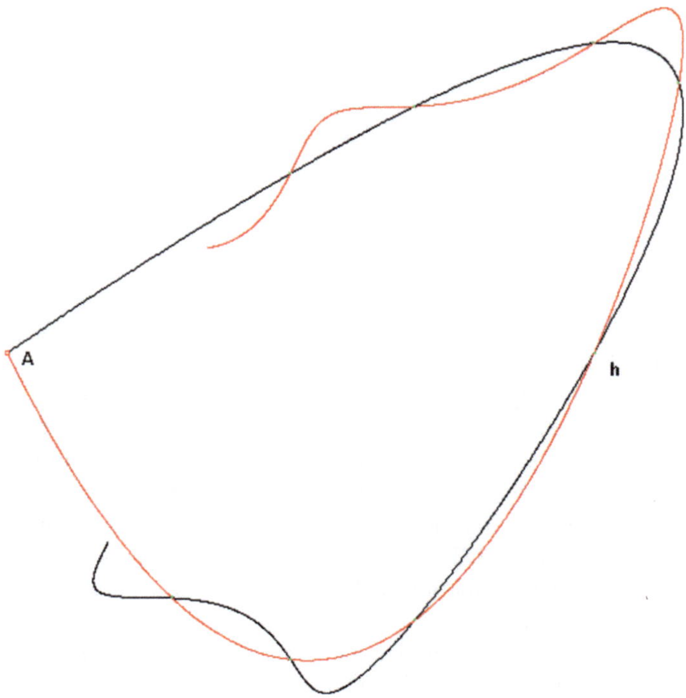

Fig. 10.9 The transverse intersection of separatrices

$$x = x_1 - y_1,$$
$$y = y_1 - a(x_1 - y_1)(1 - x_1 + y_1). \quad (10.5)$$

Location of the invariant manifolds of the mapping (10.4) is shown in Fig. 10.9.

The stable and unstable manifolds intersect at eight points. For example, the angle between the separators at the intersection point H (0.6669, 0.4169) is 0.5874 rad.

This method allows us to find parameters at which stable and unstable separatrices touch each other when they intersect. In Sect. 9.5, we consider a system with a homoclinic tangent. The mapping is

$$x_1 = bx + y + ax(1 - x)$$
$$y_1 = y + ax(1 - x). \quad (10.6)$$

The stable and unstable separatrices touch for $a = 1.511717565$ and $b = 0.93$ (Fig. 10.10). The value of the parameter a was chosen by the method of halving a segment whose ends correspond to transverse intersection of the stable and unstable manifolds and the lack of such an intersection. A similar construction for two-dimensional invariant manifolds in the three-dimensional space is considered in the book [6].

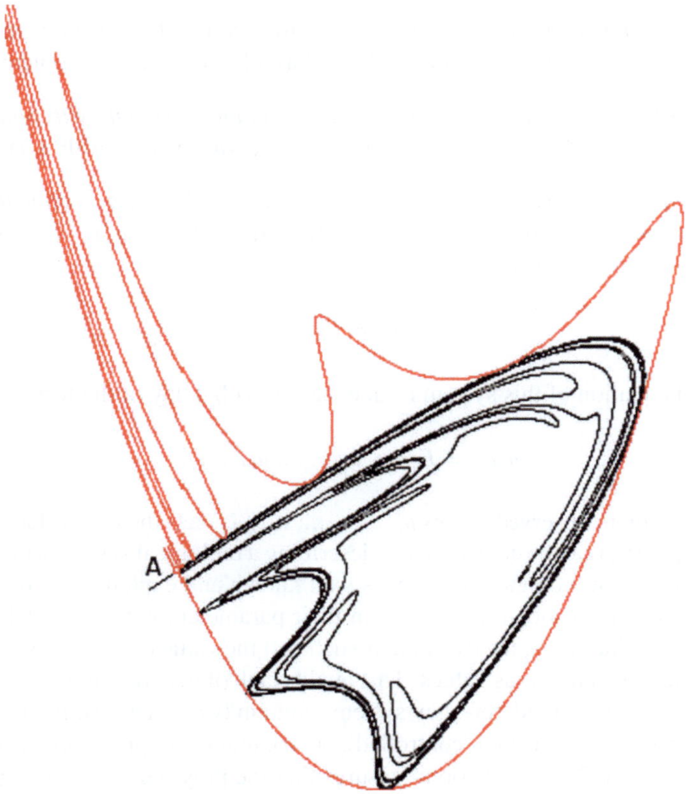

Fig. 10.10 Homoclinic tangent points for mapping (11.6)

A system with a Mobius band as an attractor is discussed in Sect. 10.4. Strip pictures 10.5 and 10.6 were created by constructing an unstable manifold of a 55-periodic hyperbolic orbit forming the boundary of a Mobius band.

10.7 Biological Systems with Memory

Let the difference equation $x_{n+1} = \Phi(x_n)$ describe the dynamics of a biological system, where $x_n \in R^d$ are the parameters of the state of the system at the n-th time period, $n \in \mathbb{Z}$. In practice, the mapping $\Phi(x)$ is determined very approximately or is often unknown. Suppose that there is an observed value $u(x_n)$ that is correctly determined every n-th period of time. For example, the mapping $\Phi(x)$ describes the dynamics of fish resources in a river–sea system. The observed value $u(x_n)$ is the number of salmon biomass that pass along the river to spawn. It is difficult to construct a dynamic system $x_{n+1} = \Phi(x_n)$, but there are methods that allow us to estimate the number of salmons $u(x_n)$ passing to spawn every year. The sequence of observations

$\{u_n, n \in \mathbb{Z}\}$ is called a time series. The goal is to describe the dynamics in terms of the observed process. Takens' theorem is the basis for solving the problem.

Theorem 10.4 [24] *Let $x_{n+1} = \Phi(x_n)$ be a dynamical system with phase space M, $dimM = d$. Let there be an observable scalar quantity $u_n = h(\Phi(x_n))$.*

Let us construct m-vectors $z_i = (u_i, u_{i+1}, ..., u_{i+m-1})$. Then in a typical case (in the case of general position) for $m \geq 2d + 1$, there exists a map $F : R^m \to R^m$ that defines the dynamic system

$$z_{i+1} = F(z_i). \tag{10.7}$$

The last equation of this system is an equation with delay of the form:

$$u_{i+m} = F_m(u_i, u_{i+1}, ..., u_{i+m-1}), \tag{10.8}$$

which, based on m observed values $u_i, u_{i+1}, ..., u_{i+m-1}$, sets the next value u_{i+m}. The delayed Eq. (10.8) can be considered as describing a biological system with memory. The delay effect in biological systems is well known and explained by the fact that biological systems respond to changes in their parameters with some delay. In the case under consideration, the dynamic is such that the value of an observed quantity is determined by m previous values. This avoids lack of general information about a system. For example, predator–prey systems contain two equations. In this case, the number of preys x is measured (controlled), but the number of predators y is not available to researcher. Takens' theorem ensures that the prey number can be predicted without knowing the predator size. To do so, it is sufficient to know current number of victims and four previous values. In each case, the required number determining values are less than $2d + 1$. Delay can be interpreted as memory of a system remembering some past dynamics. Article [8] shows that dynamics of number (mass) of biological species is naturally described by equation with delay. Obtained numerical results indicate that the dynamics of biomass is largely determined by reproduction rate, intra-specific competition, and inter-dependence between generations. External influence of habitat is modeled by perturbation of delay equations. Studied systems showed rich dynamic, where changes in parameters led to bifurcations from stable periodic regime to chaos.

Let us consider the dynamics of a biological species with a high reproduction rate. This system exhibits non-trivial chaotic behavior and has a three-dimensional structure:

$$u_{n+1} = au_n \frac{1 - bu_{n-2} - cu_{n-1} - ru_n}{1 + d(u_{n-2} + u_{n-1} + u_n)},$$

where the values are a = 175, b = 0.24, c = 0.24, r = 1, and d = 30. The variable u is a quantity that represents the ratio of the current mass to the maximum possible mass of the given biological species. Thus, u lies between zero and one. The parameters b, c, r characterize intra-specific competition and the number d characterizes the

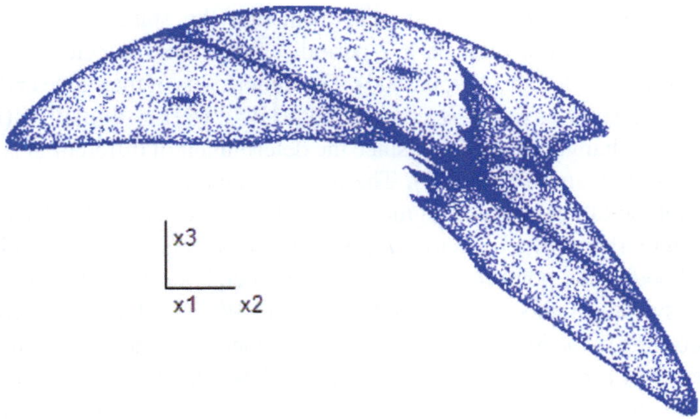

Fig. 10.11 Attractor of the system 10.9

habitat. The standard transition to the corresponding three-dimensional system leads to a system of the form:

$$\begin{aligned} x_{n+1} &= y_n \\ y_{n+1} &= z_n \\ z_{n+1} &= a z_n \frac{1 - b x_n - c y_n - r z_n}{1 + d(x_n + y_n + z_n)}. \end{aligned} \tag{10.9}$$

The system contains an attractor A, which was localized using a symbolic image (Fig. 10.11).

The structure of this attractor is formed in the following way: the system has a fixed hyperbolic point R (0.5049, 0.5049, 0.5049) (Fig. 10.12).

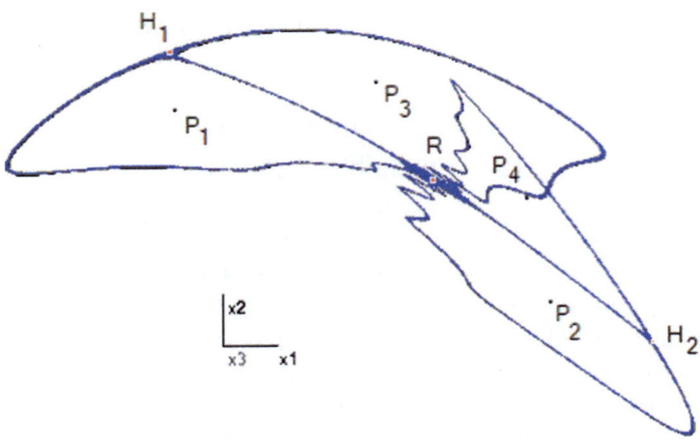

Fig. 10.12 Unstable manifold of a fixed point R

The unstable manifold is one-dimensional, and the stable manifold is two-dimensional. In addition, there is a 2-periodic hyperbolic trajectory H (0.3477, 0.6300, 0.3477), (0.6300, 0.3477, 0.6300). The dimensions of the stable and unstable manifolds of trajectories R and H coincide: $\dim\big(W^{u(s)}(R)\big) = \dim\big(W^{u(s)}(H)\big)$. The mapping (10.9) changes orientation, since the determinant of differential is less than zero, orientation changes to $W^u(R)$. The unstable manifold $W^u(R)$ is constructed using the method of iterations of segment (Sect. 10.6), see Fig. 10.12. This manifold reaches neighborhood of a trajectory H, where it intersects $W^s(H)$, then it reaches a neighborhood of the point R along $W^u(H)$. In a neighborhood of R, the manifold $W^u(R)$ transversely intersects the stable manifold $W^s(R)$, closing the homoclinic cycle. According to the Smale-Birkhoff's theorem (see [25]) there exists an invariant chaotic set Ω in a neighborhood of the manifold $W^u(R)$, the dynamics on which is equivalent to shift in binary sequence space. The set Ω is a Cantor set. Moreover, it is a component of a chain-recurrence set. The attractor A is the closure of two-dimensional unstable manifold $W^u(P)$ of 4-periodic hyperbolic trajectory P, see Figs. 10.11 and 10.12. The coordinates of 4-periodic point P_1 can be estimated as (0.3595, 0.5623, 0.4843). Inside the attractor $A = \overline{W^u(P)}$ lies another attractor $A_0 = \overline{W^u(R)}$, which is the closure of unstable manifold of point R.

10.8 Macroeconomic Dynamics

Let us consider the dynamics of the macroeconomic system "national income—interest rate—price level." This dynamic is described by the "IS-LM" model, which underlines the description of a modern market economy, see monographs [26, 27]. To model macroeconomic dynamics, we use a discrete dynamic system:

$$u_{n+1} = F(u_n), \tag{10.10}$$

where u_n describes the state of the economy at time $t = n$, $n \in \mathbb{Z}$. This system has the following form:

$$\begin{aligned} x_{n+1} &= x_n \exp\big(\mathrm{a}\big(1 - x_n y_n^{m/s} z_n\big)\big) \\ y_{n+1} &= y_n \exp\big(\mathrm{b}\big(1 - x_n y_n^{m/s} z_n\big)\big) \\ z_{n+1} &= z_n \exp(\mathrm{c}(y_n - z_n)), \end{aligned} \tag{10.11}$$

where $x = P/P_e$, P is the price level, P_e is its equilibrium value, $y = (r_e/r)^s$, r is interest rate, r_e is its equilibrium value, and $z = Y/Y_e$, Y is national income, and Y_e is its equilibrium value. All variables are dimensionless and positive quantities and their variations are regarded as deviations from equilibrium state. Parameters a, b, c, m, s are positive and $0 < s < 1$. We can say that the variable x is proportional to the price level, the variable z is proportional to the national income, and the variable y is inversely proportional to the s-th power of the interest rate. Comparing dynamics of systems for general case $m \geq 0$ and special case $m = 0$, we can notice some analogy. Indeed, they are topologically similar.

Proposition 10.1 [28]. *The mapping Φ of the form*

$$\begin{aligned} X &= xy^{m/s}, \\ Y &= y, \\ Z &= z \end{aligned} \tag{10.12}$$

displays discrete system

$$\begin{aligned} x_{n+1} &= x_n exp\big(a\big(1 - x_n y_n^{m/s} z_n\big)\big), \\ y_{n+1} &= y_n exp\big(b\big(1 - x_n y_n^{m/s} z_n\big)\big), \\ z_{n+1} &= z_n exp(c(y_n - z_n)) \end{aligned} \tag{10.13}$$

into system

$$\begin{aligned} X_{n+1} &= X_n exp(d(1 - X_n Z_n)) \\ Y_{n+1} &= Y_n exp(b(1 - X_n Z_n)) \\ Z_{n+1} &= Z_n exp(c(Y_n - Z_n)), \end{aligned} \tag{10.14}$$

where $d = a + \frac{bm}{s}$.

Comment. The mapping Φ is a diffeomorphism in the domain $\{x > 0, y > 0, z > 0\}$, where systems (10.13) and (10.14) are given. Taking into account that $x = \frac{P}{P_e}, y = (r_e/r)^s$, P is the price level and r is the interest rate then the entered coordinate is

$$X = P/P_e\big((r_e/r)^s\big)^{m/s} = \frac{P/r^m}{r_e^m/P_e}.$$

From an economic point of view, coordinate X is directly proportional to price level and inversely proportional to m-th power interest rate.

10.8.1 Dynamics Near Fixed Points

Let us consider discrete dynamical system (10.11). From Proposition 10.1, it follows that it suffices to investigate the case where $m = 0$. This somewhat simplifies calculations and reveals the essence of the dynamic. The equilibrium state is determined by the fixed points of the system (10.10), that is, by equation $F(u) = u$. Passing to the coordinates (x, y, z) and system (10.11), we obtain equalities

$$xy - 1 = 0 \text{ and } y - z = 0.$$

Thus, the fixed points fill the curve

$$K = \{xy = 1, y = z\}.$$

The projection of K onto the (xy)-plane is a hyperbola $xy = 1$, and projection of K onto the (yz)-plane is straight line $y = z$. The topological type of fixed point $u^* \in K$ of the system (10.10) is determined by eigenvalues of differential $DF(u^*)$. Invariance of curve K leads to the fact that eigenvector of differential is tangent to K. Moreover, since mapping F onto K is identical, we can conclude that $\lambda = 1$ is the eigenvalue of differential $DF|_K$. On the curve K, differential of right-hand side of system (10.11) for $m = 0$ has form:

$$D = \begin{pmatrix} 1-a & 0 & -ax^2 \\ -by^2 & 1 & -b \\ 0 & cz & 1-cz \end{pmatrix}. \tag{10.15}$$

Fixed-point multipliers are determined by equation:

$$det(D - \lambda E) = (1-\lambda)\big(\lambda^2 - \lambda(2-a-cy) + 1 - a + (b+a-1)cy\big) = 0.$$

It is clear that the multiplier $\lambda = 1$ corresponds to the curve of fixed points K. Remaining multiplicators $\lambda_{1,2}$ are given by equation:

$$\lambda^2 - \lambda(2-a-cy) + 1 - a + (b+a-1)cy = 0. \tag{10.16}$$

These yield

$$\lambda_{1,2} = \frac{2-a-cy \pm \sqrt{\Delta}}{2},$$

where discriminant

$$\Delta = c^2y^2 - 2cy(a+2b) + a^2.$$

It is easy to show that $\lambda_{1,2} \neq 1$ for $y > 0$. Therefore, eigenspace E for $\lambda_{1,2}$ is transverse to K. If $|\lambda_{1,2}| < 1$, differential contracts to K. If $|\lambda_{1,2}| > 1$, the differential stretches from K. If $|\lambda_1| < 1$ and $|\lambda_2| > 1$, near K we have hyperbolic dynamic. Discriminant Δ is negative when y lies between the roots y_1 and y_2 of equation

$$c^2y^2 - 2cy(a+2b) + a^2 = 0, \tag{10.17}$$

where

$$y_{1,2} = \frac{(a+2b) \pm 2\sqrt{b(a+b)}}{c}.$$

Thus, for $a = 2.4, b = 0.87, c = 0.9$, we have $y_1 = 0.852, y_2 = 7.51$. Therefore, for $0.852 < y < 7.51$, the multiplicators $\lambda_{1,2}$ are complex conjugate. In this case, the free term of Eq. (10.16) is the square of the modulus of the multiplicator

$$1 - a + (b + a - 1)cy = \lambda_1\lambda_2 = |\lambda|^2,$$

which makes it possible to determine domains of stable and unstable fixed points. For $a = 2.4, b = 0.87, c = 0.9$, and $y = 0.852$, we have $|\lambda|^2 = 0.3406$, that is, we have a stable focus in the plane transverse to the curve of fixed points.

Numerical experiments show that at $1 - a + (b + a - 1)cy = 1$ a Neimark–Sacker bifurcation occurs: the equilibrium loses stability and a stable invariant curve (homeomorphic to the circle) separates from it. At $a = 2.4, b = 0.87, c = 0.9$, bifurcation also occurs at $y = 1.174743$. On the curve K, the stable equilibrium states lie between the points $0.852 < y < 1.174743$. If $y > 1.174743$, then the equilibrium states are unstable. According to the Pliss reduction theorem (see [20, 29]), each fixed point on the curve K generates an invariant disk transversal to it. This disk is a stable manifold W^s for $|\lambda| < 1$ $(0.852 < y < 1.174743)$, or it is an unstable manifold W^u for $|\lambda| > 1$ $(1.174743 < y < 7.51)$. The invariant disk W^u of the point $B(0.851, 1.175, 1.175)$ is shown in Fig. 10.13. The disk W^u contains an unstable equilibrium state B and an invariant stable ellipse E. Equilibrium states with two negative multiplicators, whose absolute value is less than 1, lie between points $0.700 < y < 0.852$ on the curve K. At $0.685 < y < 0.700$ the absolute value of one multiplier is greater than 1, and the absolute value of the other multiplier is less than 1. Near $y = 0.685$, one multiplier is equal to 0, so the Jacobian of the right-hand side of system (10.11) is 0. For $y < 0.685$, one multiplier is positive and another is negative. In this case, the differential changes orientation (on the unstable manifold). The sign of Jacobian $detD$ determines whether the dynamical system preserves its orientation or not. Therefore, the equation $detD(x, y, z) = 0$ describes surface Π on which the system changes orientation. Thus, every equilibrium state in plane transverse to K can be stable, unstable with complex multipliers, or hyperbolic. In the last case, one multiplier is negative, and the second changes sign on curve Π segments $H = \{\text{hyperbolic fixed points}\}$,

$S = \{\textit{stable fixed points}\}$and

$UC = \{\textit{unstable fixed points with complex multiplicators}\}$

shown in Fig. 10.13.

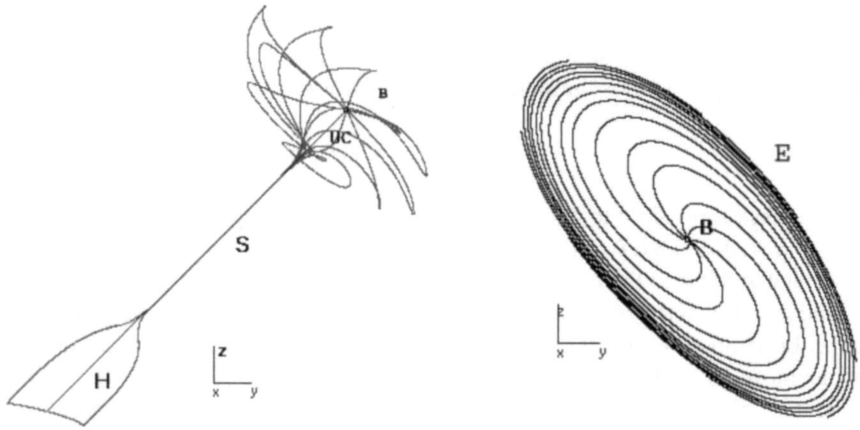

Fig. 10.13 Dynamics near the curve K of the fixed points of system (10.11)

10.8.2 Foliation with Invariant Layers

In this section, we show that there exists a function $U(x, y, z)$ whose level surfaces $U(x, y, z) = const$ are invariant for the discrete system:

$$\begin{aligned} x_{n+1} &= x_n \mathrm{exp}(\mathrm{a}(1 - x_n z_n)) \\ y_{n+1} &= y_n \mathrm{exp}(\mathrm{b}(1 - x_n z_n)) \\ z_{n=1} &= z_n \mathrm{exp}(\mathrm{c}(y_n - z_n)). \end{aligned} \tag{10.18}$$

Function $U(x, y, z)$ is an analogue of an integral for an autonomous system of differential equations. To determine the function U, we note that the first and second equations of the system (10.18) differ only in values of a and b. Raising the first equation to the power b and the second to the power a, we get same expression $exp(ab(1 - x_n z_n))$. This enables us to construct function U.

Preposition 10.2 [28]. *Level surfaces of function*

$$U = \frac{x^b}{y^a}$$

are invariant for system (10.18).

Each surface

$$W = \left\{(x, y, z) : x = h y^{\frac{a}{b}}\right\}, \quad h = const$$

is transverse to curve of fixed points K. In addition, as shown above, invariant disks $W^{s,u}$ exist in neighborhood of K. Due to invariance, disks $W^{s,u}$ must lie in the layers of

W. Thus, the invariant foliation generated by stable W^s and unstable W^u disks exists not only near the curve K, but also defined globally. Since function U does not depend explicitly on z, invariant layer $W = \{(x, y, z) : x = hy^{\frac{a}{b}}, h = const\}$ is a surface with straight lines parallel to Z-axis. System on the surface $W = \{(x, y, z) : x = hy^{\frac{a}{b}}\}$ is defined as

$$\begin{aligned} y_{n+1} &= y_n \exp\left(\mathrm{b}\left(1 - hy_n^{a/b} z_n\right)\right), \\ z_{n+1} &= z_n \exp(c(y_n - z_n)), \end{aligned} \tag{10.19}$$

where number $h > 0$ defines invariant layer. For example, $h = 1$ defines a surface passing through equilibrium state $(1, 1, 1)$. For arbitrary surface $W(h)$, the equilibrium state is defined by the equalities:

$$xy = 1, \quad y = z, \quad x = hy^{a/b}.$$

This implies that equilibrium state has the coordinates

$$\left(h^{b/(a+b)}, h^{-b/(a+b)}, h^{-b/(a+b)}\right)$$

on each layer $W(h)$. Let us study changes in the dynamics of the system in the layers $W(h) = \{(x, y, z) : x = hy^{a/b}\}$ depending on value of h. In other words, we consider bifurcations of the system (10.19) caused by variation parameter h. There is a fixed point $W(h) \cap K$ of system (10.19) in each layer $W(h)$. For a fixed layer, denote the fixed point by K^*. Consider system (10.19) with $a = 2.4$; $b = 0.9$; $c = 0.9$. In layer $W(h = 1)$ there is stable equilibrium state $K^*(1, 1)$ with complex multipliers. As h decreases, character of stability begins change. On layer $W(h = 0.8)$ (Fig. 10.14) there is 3-periodic invariant set $R = \{R_1, R_2, R_3\}$, and trajectories from this set go to stable equilibrium state K^*.

When parameter h changes from 0.8 to 0.58, a stable invariant ellipse A is born from the stable equilibrium state, and equilibrium state K^* loses stability, that is, a Neimark–Sacker bifurcation occurs (Fig. 10.15).

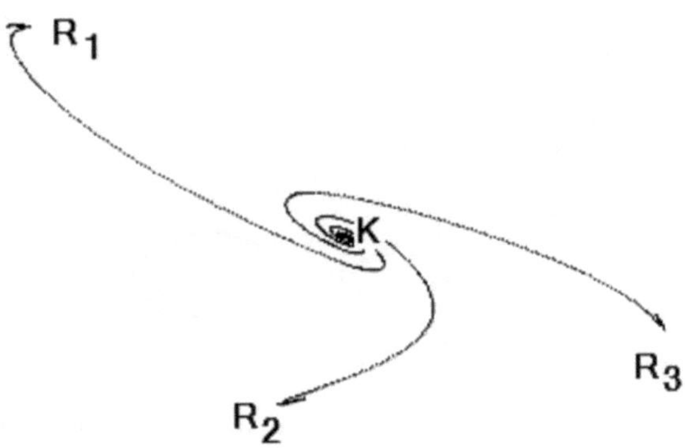

Fig. 10.14 Stable equilibrium state at h = 0.8

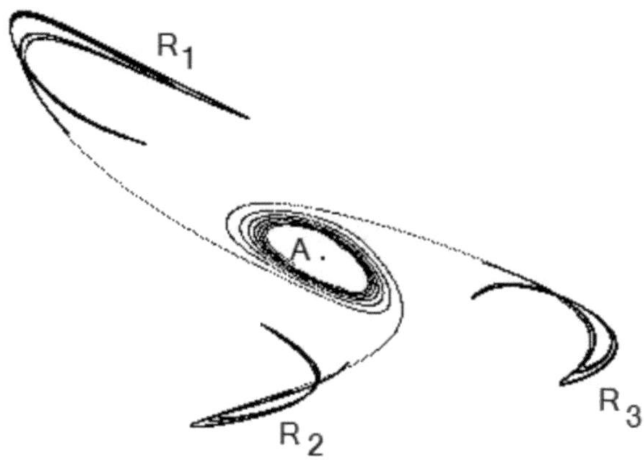

Fig. 10.15 Birth of a stable invariant ellipse with changing h from 0.8 to 0.58

In this case, the invariant set R increases. Subsequent decrease in the parameter h is accompanied birth of hyperbolic 3-periodic trajectory P_3 on the stable ellipse A (Fig. 10.16).

In the case where h reaches value 0.515970, the unstable manifold $W^u(P_3)$ intersects transversely with the stable manifold $W_s(P_1)$ of the trajectory of point $P(0.8037147\ 1.33638665)$ (Fig. 10.16). These manifolds are constructed and their transversality is estimated using method described in book [6]. According to Smale's theorem (see [25]), a transverse intersection leads to chaotic behavior near the intersection points. Moreover, the set A loses stability and merges with the set R, forming an invariant set Ω. This set is the closure of the unstable manifold $W^u(P)$ of the trajectory P (Fig. 10.16). The trajectory starting near the equilibrium state

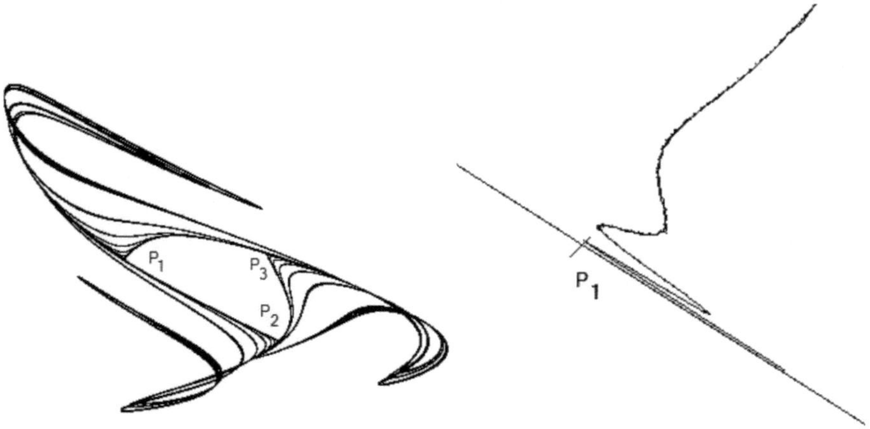

Fig. 10.16 Chaos in the invariant layer $W(h = 0.515970)$, and the closure of the unstable manifold $W^u(P)$. Intersection of unstable and stable manifolds at the 3-periodic point P

$K^*(1.1978, 1.1978)$ reaches the set Ω and then wanders around it. We can estimate the entropy on this invariant set Ω using the growth of the length of a curve (see [30]). The lower estimate for the entropy is E = 0.69314. Since entropy measures chaos, we conclude that the system exhibits chaotic behavior on this set.

10.8.3 Uncontrollable Perturbations

The macroeconomic environment influences the dynamics of the system. External influence can be considered as an uncontrollable disturbance of the system. To simulate external influences, we assume that perturbation is randomly distributed over time, it is small and can be added to the results as a whole. To simulate the perturbation, we use dependence

$$q(n+1) = 1 - 2q^2(n), q \in [-1, 1].$$

It is known (see [31]) that, for almost all initial values $q(0)$, the trajectory {q(n)} is chaotic and distributed on the segment [–1,1] with density

$$\rho = \frac{1}{\pi\left(1 - x^2\right)^{1/2}}.$$

Thus, we have a system of equations of the form:

$$\begin{aligned} x_{n+1} &= x_n \exp\left(\mathrm{a}\left(1 - x_n \mathrm{y}_n^{m/s} \mathrm{z}_n\right)\right) + \varepsilon_1 q_1(n), \\ y_{n+1} &= y_n \exp\left(\mathrm{b}\left(1 - x_n \mathrm{y}_n^{m/s} \mathrm{z}_n\right)\right) + \varepsilon_2 q_2(n), \\ z_{n+1} &= z_n \exp(\mathrm{c}(y_n - z_n)) + \varepsilon_3 q_3(n), \end{aligned} \tag{10.20}$$

where ε_i are small positive numbers, q_i takes random values from the segment [-1,1]; depends chaotically over time n. It is expected that the described perturbations do not conserve the invariant foliation. However, each invariant layer of the unperturbed system has attractors with specific domains of attraction which must be conserved under small perturbations. The results described above indicate that attractors form from stability in these equilibrium states. In this case, attractors arise as a result of loss of stability of these equilibrium states. In this case, these attractors continually change from layer to layer. The union of these attractor sets does not vanish under perturbations. This is well illustrated (Fig. 10.17) by the trajectory of point (1, 1, 1) in a perturbed system with parameters $a = 2.45, b = 0.6, c = 0.9, m = 0$, and chaotic perturbation $\varepsilon_i = 0.01$.

It should be noted that under the influence of chaotic perturbations, the trajectories move up and down near the attractors of the unperturbed system. In this case, perturbations are able not only to move at trajectory into a chaotic region (Fig. 10.17), but also to return it from that region. From an economical point of view, some

Fig. 10.17 Trajectory of the point (1, 1, 1) of the perturbed system (10.20) with $\varepsilon_i = 0.01$, and the trajectory of the same point as $\varepsilon_1 = 0.01$, $\varepsilon_{2,3} = 0$

perturbations can be reduced to zero or negligible, while others cannot be significantly decreased. For example, a central bank may control interest rates and prevent chaotic perturbations. However, it is impossible to eliminate all chaotic perturbations of price levels. Therefore, it makes sense to distinguish between perturbations whose influence on system dynamics and those that can be neglected. Note that perturbing the last equation in the system maintains invariant layers, since each layer is a ruled surface parallel to the Z-axis. Thus, a perturbation of the form $\varepsilon_1 = \varepsilon_2 = 0$ and $\varepsilon_3 \neq 0$ maintains the layers of an unperturbed system and perturbs the system within a layer. This implies that perturbations to national income (production) have little effect on dynamics near attractor in layers.

Weak price level control leads to a perturbation in the first equation. The trajectory of the point (1, 1, 1) in the perturbed system with $\varepsilon_1 = 0.01$, $\varepsilon_{2,3} = 0$ is shown in Fig. 10.17. Numerical results show that price level perturbations lead to bifurcation in y and z. An increase in y causes the interest rate to decrease. It is worth noting that the strong bifurcations toward chaotic state are observed with weak interest rate control. Numerical experiments for $\varepsilon_2 = 0.01$ and $\varepsilon_1 = \varepsilon_3 = 0$ show that this behavior of the system does not differ from that of a general type perturbed with $\varepsilon_i = 0.01$ (Fig. 10.17). Under small general type perturbations, trajectories begin to shift toward equilibrium states and then fall into an unstable equilibrium state before entering a chaotic layer. The level of chaos can increase or decrease under perturbations.

10.8.4 Conclusion

A study of a discrete macroeconomic model (10.11) was carried out. It was shown that this system is C^1-equivalent to the system (10.11) with $m = 0$. It has a curve filled with equilibrium states, and invariant level surfaces of the function $U = x^b/y^a$, forming a foliation transversal to this curve. It can be assumed that equilibrium states are the centers of each invariant surface. Each layer has an attractor, and almost all trajectories approach this attractor. This attractor can be either a state of equilibrium or have a rather complex (chaotic) structure. When the level surface varies, we observe a bifurcation in the system dynamics from stable equilibrium to chaos. Note that chaos in this macroeconomic model is an inherent property of system. Moreover, chaos does not always lead to an economic crisis. In the analyzed case, chaos means the impossibility of long-term forecasts. An economic crisis is chaos of uncontrolled deviation of significant proportions. Numerical results show that there are layers with high level of chaos, which leads to an imbalance in economic system and onset of crisis. Small external perturbations can disrupt the described topological structure of trajectories. Numerical experiments and economic practice show that not all perturbations affect the system's dynamics equally. Thus, perturbations in national income (z) do not influence invariant foliation; perturbations of price levels (x) lead to minor changes in dynamics while preserving attractors in layers. At the same time, fluctuations in interest rates greatly affect system dynamics. With small perturbations, orbit begins to move through equilibria and falls first into unstable equilibrium state, then into layer of chaos whose magnitude can increase and become very large.

References

1. Avrutin, V., Levi, P., Schanz, V., Fundinger, D., Osipenko, G.: Investigation of dynamical systems using symbolic images: efficient implementation and applications. Int. J. Bifurc. Chaos **16**(12), 3451–3496 (2006)
2. Fundinger, D.: Investigating Dynamics by Multilevel Phase Space Discretization. PhD thesis, Stuttgart University (2006)
3. Petrenko, E.I.: Computer research of dynamic systems based on the symbolic image method. Ph.D. dissertation, St. Petersburg University (2009)
4. Pissanetzky, S.: Sparse Matrix Technology. Academic Press, New York (1984)
5. Tarjan, R.: Depth-first search and linear graph algorithms. SIAM J. Comput. **1**, 146–160 (1972)
6. Osipenko, G.: Dynamical Systems, Graphs, and Algorithms. Lectures Notes in Mathematics, 1889, Springer, Berlin (2007)
7. Fundinger, D., Lindstrem, T., Osipenko, G.S.: On the appearance of multiple attractors in discrete food-chains. Appl. Math. Comput. **184**(2), 429–444 (2007)
8. Osipenko, G.S.: Structural Stability and Symbolic Image. Differential Equations, Pleiades Publishing, Inc. Original Russian Text published in Differentsial'nye Uravneniya **42**(4), 1–9 and 468–475 (2006)
9. Gragnami, A., De Feo, O., Rinaldi, S.: Food chains in the chemostats: relation between mean yield and complex dynamics. Bull. Math. Biol. **60**, 703–719 (1998)
10. Rosenzweig, M.L., MacArthur, R.H.: Graphical representation and stability conditions of predator-prey interactions. Am. Nat. **97**, 209–223 (1963)
11. Rosenzweig, M.L.: Exploitation in three trophic levels. Am. Nat. **107**, 275–294 (1973)
12. Lindstrm, T.: On the dynamics of discrete food chains. Low- and high-frequency behavior and optimality of chaos. J. Math. Biol. **45**, 396–418 (2002)
13. Cochet-Terrason, J., Cohen, G., Gaubert, S., McGettrick, M., Quadrat, J.-P.: Numerical computation of spectral elements in max-plus algebra. IFAC Conference on System Structure and Control (1998)

14. Cormen, T.H., Leiserson, C.E., Rivest, R.L., Stein, C.: Introduction to Algorithms. MIT Press, Cambridge (2002)
15. Karp, R.M.: Characterization of the minimum mean-cycle in a digraph. Discret. Math. **23**, 309–311 (1978)
16. Romanovsky, J.V.: Optimization and stationary control of discrete deterministic process in dynamic programming. Kibernetika, **2**, 66–78 (in Russian); Engl. transl.: Cybernetics, **3** (1967).
17. Romanovsky, J.V.: Algorithms for Solving Extremal Problems. Nauka, Moscow (1977). (in Russian)
18. Siek, J.G., Lee, L.Q., Lumsdaine, A.: C++ Boost Graph Library. Addison-Wesley, Boston—San Francisco—New York (2001)
19. Osipenko, G.S., Romanovsky, J.V., Ampilova, N.B., Petrenko, E.I.: Computation of Morse spectrum. J. Math. Sci. **120**(2), 1155–1166 (2004)
20. Pliss, V.A.: Integral Sets of Periodic System of Differential Equations. Moscow, 1977 (in Russian)
21. Poincaré, H.: Sur le problème des trois corps et les équations de la dynamique. Acta Math. **13**, 1–271 (1890)
22. Poincaré, H.: Sur une forme nouvelle des équations de la Mécanique. Comptes Rendus de l'Academie des Sciences. **132**, 369–371 (1901)
23. Smale, S.: Diffeomorphisms with many periodic points. Differential and Combinatorial Topology. Princeton University Press, pp. 63–80 (1965
24. Takens, F.: Detecting strange attractor in turbulence. Lect. Notes in Math., Springer, Berlin **898**, 89–94 (1981)
25. Katok, A., Hasselblatt, B.: Introduction to the Modern Theory of Dynamical Systems. Cambridge University Press (1995)
26. Pesenti, A.: Manual of Political Economia, vol. 1, Riuniti, Roma (1970)
27. Stoleru, L.: L'équilibre et la croissance économiques. Principes de macroéconomie. Dunod, Paris (1969)
28. Osipenko, G.S.: Nonlinear dynamics of interaction of national income, interest rate, and price level. J. Math. Sci. **205**(6), 806–820 (2015)
29. Osipenko, G.: Center Manifolds. In: Meyers, R.A. (ed.) Encyclopedia of Complexity and Systems Science. Springer, NY, Berlin, pp. 936–951 (2009)
30. Newhouse, S., Pignataro, T.: On the estimation of topological entropy. J. Stat. Phys. **72**, 1331–1351 (1993)
31. Schuster, H.G., Just, W.: Deterministic Chaos: An Introduction. Wiley-VCH, Weinheim (2005)

Chapter 11
Appendix

11.1 Measure and Topology

Measurable sets and measure. For more information on this topic, see [1, 2].

Definition 11.1 Consider a set M. A nonempty collection Σ of its subsets is called a σ-algebra if the following conditions are met:

(1) *the set M belongs to the collection Σ;*
(2) *if a set A lies in the collection Σ, then its complement $M \setminus A$ also lies in it;*
(3) *the union and intersection of countable family of sets from Σ belong to Σ.*

The sets in the algebra Σ are called measurable and the measure of these sets is defined as follows.

A measure is a non-negative function μ defined on Σ that satisfies the following properties:

(1) $\mu(\emptyset) = 0$,
(2) if $A \subset B$, then $\mu(A) \leq \mu(B)$,
(3) the function μ is σ-additive, that is, the measure of at most a countable number of disjoint union of sets is equal to the sum of the measures of these sets:

$$\mu\left(\bigcup_i A_i\right) = \sum_i \mu(A_i),$$

where the sets A_i are pairwise disjoint, and the number of sets can be finite or countable. Since the measure of any set must be non-negative, the right-hand row converges to some non-negative value or to $+\infty$. We consider finite measures where the right-hand side converges to a non-negative value.

A measure has the following property of continuity:

G. Osipenko, *Symbolic Analysis of Dynamical Systems*, Mathematical Marvels: Texts and Monographs in the Spirit of CR Rao, https://doi.org/10.1007/978-981-95-0713-9_11

(1) for a sequence of measurable nested sets

$$A_1 \supseteq A_2 \supseteq A_3 \supseteq \cdots \text{the set A} = \bigcap_{n=1}^{\infty} A_n$$

is measurable, and the measure

$$\mu(A) = \lim_{n\to\infty} \mu(A_n);$$

(2) for an expanding sequence of measurable sets

$$A_1 \subseteq A_2 \subseteq A_3 \subseteq \cdots \text{the set A} = \bigcap_{n=1}^{\infty} A_n$$

is measurable, and the measure

$$\mu(A) = \lim_{n\to\infty} \mu(A_n)$$

A measure space is a triple $\{M, \Sigma, \mu\}$ consisting of a set, a σ-algebra, and a measure. The measure space, where $\mu(M) = 1$, is called a probability space, and the measure μ is called normalized. If measure of the ambient space $\mu(M) = m$ is finite, then the measure $\mu^* = \frac{\mu}{m}$ is normalized. If the manifold M is compact, then without loss of generality we can assume that the measure under consideration is normalized. Assume the set M has a topological structure, that is, open and closed sets are defined. We say that a σ-algebra is compatible with the topological structure, if it is generated from closed subsets of M. In this case, Σ is called the Borel σ-algebra, and measure μ is the Borel measure. The Borel measure has a regularity property: for each measurable set $A \in \Sigma$

$$\begin{aligned}\mu(A) &= \inf\{\mu(B), A \subset B,\ B \text{ is open}\}\\ &= sup\{\mu(K), K \subset A, K \text{ is compact}\}.\end{aligned}$$

Lebesgue Integral. Suppose M is a manifold with a Borel σ-algebra Σ and a measure μ. A function $\varphi : M \to R$ is said to be measurable if the set $\{x : \varphi(x) > \alpha\}$ is measurable for all $\alpha \in R$. If function φ is measurable, then the sets $\{x : \alpha < \varphi(x) < \beta\}$ and $\{x : \varphi(x) = \alpha\}$ are measurable, where α and β are any numbers. Let the function φ be bounded, that is, there exist p and q such that $p \le \varphi(x) \le q$. Let us divide the interval $[p, q]$ by intermediate points $l_0 = p < l_1 < l_2 < \cdots < l_n = q$ and form an integral sum:

$$\sum_{k=0}^{n} l_k \mu(E_k) + \sum_{k=0}^{n} l_k \mu(E_k^*),$$

where $E_k = \{x : l_k < \varphi(x) < l_{k+1}\}$ and $E_k^* = \{x : \varphi(x) = l_k\}$. It can be shown that if the largest of $l_{k+1} - l_k$ tends to 0, then there exists a limit for the integral sums, which is called the Lebesgue integral of the function φ with respect to the measure μ and denoted by

$$\int_M \varphi d\mu = \lim_{\tau \to 0} \left(\sum_{k=0}^{n} l_k \mu(E_k) + \sum_{k=0}^{n} l_k \mu(E_k^*) \right),$$

where $\tau = \max|l_{k+1} - l_k|$. If this limit is finite then φ is called summable. If function $\varphi \geq 0$ is unbounded, then the function

$$\varphi_n(x) = \begin{cases} \varphi(x), & \varphi(x) \leq n \\ n, & \varphi(x) > n. \end{cases}$$

Then by definition

$$\int_M \varphi d\mu = \lim_{n \to \infty} \int_M \varphi_n d\mu.$$

If this limit is finite, then the function φ is called summable. The integral of a negative measurable function is defined similarly. Finally, any measurable function can be represented as the sum

$$\varphi = \varphi^+ + \varphi^-,$$

where

$$\varphi^+(x) = \begin{cases} \varphi(x), & \varphi(x) > 0, \\ 0 & \varphi(x) \leq 0, \end{cases}$$

$$\varphi^-(x) = \begin{cases} \varphi(x), & \varphi(x) < 0, \\ 0 & \varphi(x) \geq 0. \end{cases}$$

Then

$$\int_M \varphi d\mu = \int_M \varphi^+ d\mu + \int_M \varphi^- d\mu,$$

if both integrals on the right-hand side are finite. In this case, function φ is called μ-summable or integrable.

We will say that a measure μ is absolutely continuous with respect to another measure σ if there exists a measurable function φ such that for any measurable set A

$$\mu(A) = \int_A \varphi d\sigma.$$

Space of summable functions L_1 [3].

Let us consider the space $\{M, \Sigma, \mu\}$ with a normalized measure and a set of integrable functions. Further, all integrals are taken over the same set M; so we can omit it from the integrals. If the function φ is integrable, then there exists $\int |\varphi| d\mu$ and this value can be taken as a norm. We obtain a Banach space of integrable functions, which is denoted by L_1 or $L_1(M, \mu)$, if we need to specify the space and (or) measure.

The space of continuous functions is dense in L_1.

The characteristic function or indicator of set A is defined as

$$\chi_A(x) = \begin{cases} 1, & as\ x \in A \\ 0, & as\ x \notin A. \end{cases}$$

For any compact set K and any open neighborhood $U \supset$ K, there exists a continuous function $\varphi : M \to [0, 1]$ such that

$$\chi_K(x) \leq \varphi(x) \leq \chi_U(x).$$

A nested sequence of neighborhoods U_n, contracting to K, generates a sequence of continuous functions φ_n pointwise converging to the characteristic function χ_K.

11.2 Graphs

Terminology. A directed graph G is a set of vertices (nodes) $V = \{i, j, r, \cdots\}$ and a set of oriented edges (arcs) $E = \{i \to j, k \to i, \cdots\}$. Loops $i \to i$ are allowed. Two edges with a common starting vertex $v \to u_1$ and $v \to u_2$ are called adjacent. If there is an arc $v \to u$, then we say: v is the ancestor of u, and u is a descendant of v, while the vertices u and v are adjacent. The graph G has two mappings: b, e from the set of arcs to the set of vertices that define the beginning and ending of the arc:

$$b(i \to j) = i, e(i \to j) = j.$$

The sequence $\omega = \{i_1, i_2, \cdots, i_m\}$ is a path on G if for every n there exists an edge $i_n \to i_{n+1}$, while i_1 is the beginning and i_m is the end of that path. If there exists a path ω that starts at the vertex v and ends at the vertex u, then the vertex u is called reachable from v, the symbol $\longmapsto$ will denote reachability. The sequence $\omega^* = \{i_1, i_2, \cdots, i_m\}$ is a chain on G if for every n there exists an edge of the form $i_n \to i_{n+1}$ or of the form $i_{n+1} \to i_n$, that is, the chain does not take into account the orientation of the arcs. The chain ω^* is closed if in $i_1 = i_m$. A connected component is the maximal inclusion subset of the vertices $H \subset G$ when any two vertices are connected in a chain. Obviously, the graph G decomposes into connected components. A connected graph (component) is a tree if there are no closed chains. The following theorem is proved in the monograph [4].

Theorem 11.1 *The following definitions of trees are equivalent.*

(1) *A connected graph without closed chains.*
(2) *A graph without closed chains, the number of arcs of which is one less than the number of vertices.*
(3) *A connected graph whose number of arcs is one less than the number of vertices.*
(4) *A graph without closed chains, in which adding an arc leads to the appearance of a closed chain.*
(5) *A connected graph that is no longer connected after removing any arc.*
(6) *A graph in which any two vertices are connected by a chain and only one chain.*

A strong connected component is a subset of vertices $H \subset G$ that is maximal in inclusion when any two vertices in this subset are reachable from each other:

$$u, v \in H \ : \ u \longmapsto v, v \longmapsto u.$$

Computer representation of a graph. From a theoretical point of view, it is convenient to represent a directed graph G by its (adjacency) matrix of admissible transitions

$$\Pi = \left(\pi_{ij}\right),$$

where $\pi_{ij} = 1$ if there exists an edge $i \to j$, otherwise $\pi_{ij} = 0$. In practice, the transition matrix is large and sparse. That is, it contains many 0's and few 1's. It is inconvenient to work with such matrices on a computer. Therefore, a graph is usually represented as a collection of adjacency lists. For each vertex u, a list of vertices adjacent to the given one is created. This is equivalent to listing the arcs outgoing from the vertex u. For example, an adjacency list $S(u) = (u; i_1, i_2, \cdots)$ means that the graph G has a vertex u from which the edges $u \to i_1, u \to i_2, \cdots$ go out. In this case, the number of elements in the list $S(u)$ is greater by one more than the number of 1's in the corresponding row of the matrix Π. The graph G is given by enumeration of adjacency lists

$$S(u_1),$$

$$S(u_2),$$

The resulting list of lists is a fairly compact definition of a graph with a large number of vertices. Representing a graph in this way uses memory of the size O(n + m), where n is the number of vertices, m is the number of edges.

11.3 Depth-First Graph Traversal

The goal of the algorithm is to optimize the traverse all vertices in the graph along directed arcs, while not using all arcs. From the name of this method, it can be seen that during the search process, we go as deep into the graph as possible. Following the algorithm, we traverse all vertices accessible from the starting vertex sequentially. If an edge leads to a previously unvisited vertex we follow that edge. If there are no further edges leading to unexplored vertex, a return occurs. To determine the state of a vertex, colors are often used: white, gray, black. The white vertex indicates that it has not been considered, meaning that all vertices are initially white. The gray vertex means it is working, has been visited, and the exploration of all outgoing edges has not been completed. A gray vertex is on a downward path in a tree. The black color of the vertex indicates that the exploration is completed, all outgoing edges have been examined, and we should return to the previous if there is one. The starting vertex of the depth-first traversal is called the "root" vertex. If the root vertex is black, we have built a maximal traversal tree from that vertex. Note that, if the root is the starting vertex, the resulting deep-first tree includes vertices reachable from it. If the graph has white vertices, we must continue traversing from one of them. Thus, the traversal stops when all vertices become black.

Example of Depth-First Search on Fig. 11.1.

1. Assign the white color to all vertices. Check that the first vertex is white. Go to it and make it gray.
2. Try going to the vertex 2. Check it is white and go there. Make it gray.
3. Try to reach vertex number 3. Check that it is white and go to it. We paint it gray.
4. Make sure that no edges coming from the vertex number 3. We mark it black and return to the vertex numbered 2.
5. Try reaching vertex number 4. Check that it is white, and go there, paint it gray, make sure there are no edges leading from vertex number 4, mark it black, and return to vertex number 2.
6. There are no more outgoing edges from the vertex number 4. Mark it black and return to vertex number 2.
7. No more edges from vertex number 2. We mark it black and return to vertex number 1.
8. There are no more outgoing edges from the vertex number 1. Mark it black and check that all vertices are black. The algorithm is complete.

The algorithm did not use the edges $4 \to 3$ and $4 \to 1$, if we remove them from the graph, then we get a tree of depth-first traversal. Resulting tree is a subgraph of the original graph. Traversal tree depends on the initial vertex choice. Start traversing from vertex 4 gets another tree. If the original graph is rich enough, then we get a set of trees, which is called a depth-first traversal forest. Convenient to use timestamps instead of vertex colors.

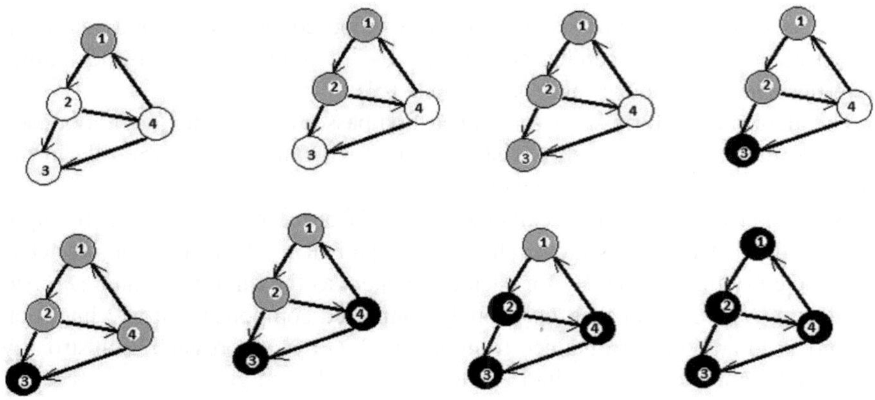

Fig. 11.1 Bypass of the graph in depth

Traversal with timestamps

Depth-first traversal of a graph with timestamps is different from the traversal described above in that instead of coloring, some numerical characteristics of the vertex are introduced. For each vertex u, specify two numbers: input "time" — $entry[u]$ and exit "time" — $leave[u]$. Let us modify the procedure described above as follows. First, introduce a time counter t and set $t = 0$. Let there be an unconsidered (white) vertex u and for it, describe the $DFS(u)$ procedure: increase the current time t by 1 and set the entry time $entry[u] = t$ (instead of painting the vertex gray). For any vertex v adjacent to u that does not have a labeled $entry[u]$ (painted white), we perform the DFS (v) procedure. If there are no more edges leading to the unconsidered vertex, backtracking occurs. If all neighboring vertices of vertex u are marked with $entry[v]$, then increase the current time by 1 and set the exit time $leave[u] = t$ (instead of repainting vertex u black). Obviously, for each vertex $entry[u] < leave[u]$. If we finish building a tree with the root vertex u, the exit time is $leave[u]$ and go to a new unconsidered (white) vertex v, the time counting continues, and the vertex v receives the entry time $entry[v] = leave[u] + 1$. Thus, after completion of the traversal each vertex receives two numbers: $entry[u]$ and $leave[u]$.

11.4 Strongly Connected Component

Strongly connected component H is a collection of vertices in a graph such that there is a closed path between every pair of vertices from H:

$$H = \{u, v \in H \ : u \longmapsto v, v \longmapsto u\}.$$

It follows that from any vertex $u \in H$ all vertices of the set H are reached.

The following algorithm was independently proposed by Kosaraju and Sharir [5]. It is based on two depth-searches and runs in $O(n + m)$ time, where n is the number of vertices, m is the number of arcs (see book [6]).

The algorithm performs a series of deep-first passes through the graph. For each vertex u, the exit time $leave[u]$ is stored. This exit time plays a key role in the algorithm and is expressed in the theorem below.

We introduce a new notation: exit time $leave[H]$ from a strongly connected component H is defined as the maximum values $leave[u]$ for all $u \in H$. In addition, we determine the entry times $entry[H]$ for each component as the minimum value $entry[u]$ for all $u \in H$. It is clear that the strongly connected components do not intersect. A condensed graph G^* is obtained from the graph G by compressing each strongly connected component into a single vertex. The condensed graph has two types of vertices:

(1) a vertex i corresponding to a strongly connected component H_i,
(2) a vertex i representing a nonrecurrent (passing) vertex in the graph G, in which case, we assume $\{i\} = H_i$.

Directed edges between two vertices i and j of the condensation graph G^* exist only between different vertices. This eliminates the loop $i \to i$. An edge $i \to j$ exists on G^* if there is an edge on G from some element of H_i to some element of H_j. According to the construction, the condensed graph G^* has only nonrecurrent vertices. In this case, the graph is called acyclic.

Theorem 11.2 *Let i and j be two different vertices in the condensation graph. Let there be a path* $i \longmapsto j$. *Then* $leave[H_i] > leave\left[H_j\right]$.

Proof. When traversing the graph in depth-first, two fundamentally different scenarios arise depending on which set of vertices H_i or H_j is visited first.

1. Let H_i be visited before H_j. That is $entry[H_i] < entry\left[H_j\right]$. This means that at the moment $t = entry[H_i]$ the traversal reaches a vertex $v \in H_i$, while all other vertices of the sets H_i and H_j have not yet been visited. By assumption, there exists a path $H_i \longmapsto H_j$ in the condensed graph G^*. This means the existence of the path $w \longmapsto u$, where $w \in H_i$, $u \in H_j$. If H_i is a component, $v \in H_i$, $w \in H_i$, then either $v = w$ or there is a path $v \longmapsto w$. Then the component H_i and the component H_j are reachable from the vertex v. This means that traversal from the vertex v goes through all the vertices of the components H_i and H_j, that is, for any vertex $u \in H_i \cup H_j$, $u \neq v$ $leave[v] > leave[u]$.
2. Let component H_j be achieved first, that is, $entry\left[H_j\right] < entry[H_i]$. This means that at the moment $t = entry\left[H_j\right]$, the traversal reaches some vertex $v \in H_j$, and all the other vertices of the components H_i and H_j are not visited. Since, by hypothesis, the path $i \longmapsto j$ existed in the condensation graph and due to the acyclicity of the condensed graph, there is no return path $j \longmapsto i$, that is, the deep passage from the vertex v does not reach the vertices of H_i. This means that they are visited later during the depth-first traverse, it follows $leave[H_i] > leave\left[H_j\right]$. The theorem is proven.

The proven theorem is the basis for the algorithm for finding strongly connected components. Any path $i \longmapsto j$ in the condensation graph goes from a component with a larger value *leave* to a component with a smaller value. Thus, if we sort all the vertices v in descending order by exit time $leave[v]$, then the first one will be a vertex u with the maximum value $leave[v]$ belonging to the "root" of a strongly connected component of $H(u)$ that has no exit. To construct such a strong component $H(u)$ consider the transposed graph G^T, that is, the graph obtained from G by reversing the directions of each edge. It is easy to see that this graph has the same components of strong connection as the original graph. Moreover, its condensation graph $\left(G^T\right)^*$ is equal to the transposed condensation graph $(G^*)^T$ of the original graph. This means that G^T has no exit from the component of $H(u)$, that is, it has an input but no output. Thus, in order to bypass the entire strongly connected component $H(u)$ containing some vertex u, it is enough to start a traverse from the vertex u in the graph G^T. Then this tour will visit all the vertices of this strong connected component and only them. Next, we can remove the vertices of H from the graph G along with the edges outgoing from H. Then we find the next vertex v with the maximum value $leave[v]$. This vertex can be either recurrent or nonrecurent. A vertex is nonrecurrent, if only it can be reached from the "root" of the strong component H. Then, after removing H together with the outgoing edges, v becomes nonrecurrent if there are no edges of the form $w \to v$. In this case, the vertex v should be removed from the graph G together with the edges leaving it. If v is recurrent, then we start with it a deep traversal on the transposed graph. Thus, the following algorithm for the selection of strongly connected components is obtained:

(1) Go around the depth of the graph G and arrange the vertices in decreasing order of the exit time $leave[v]$.
(2) Construct the transposed graph G^T.
(3) Determine vertex $v \in G$ with the maximum value $leave[v]$.
(4) Check for an edge of the form $w \to v$ with a maximum value of $leave[v]$. (This means that v is a recurrent vertex.) If there is such an edge, go to step 5; otherwise, set $H(v) = \{v\}$ and go to step 6.
(5) Start a depth-first traversal on the graph G^T from the vertex v to obtain component $H(v)$ and add it to a list of components.
(6) Remove the vertices $H(v)$ from G and G^T together with the edges outgoing from them. If $G \neq \emptyset$ (that is, there are no removed vertices), then go to step 3, otherwise the algorithm is completed.

Note that removing vertices and their outgoing edges can be done by labels u^*.

11.5 Canonical Sorting

Let us consider a directed graph G. Let us introduce a quasi-order relation between the vertices of the graph G. We write $i \longmapsto j$ if and only if there exists an admissible path of the form

$$i = i_0, i_1, i_2, ..., i_m = j.$$

A vertex i is recurrent if and only if $i \longmapsto i$, and a pair of recurrent vertices i, j is equivalent if $i \longmapsto j \longmapsto i$, that is, there is a common closed path through i and j. The matrix of admissible transitions Π depends on the numbering of the vertices, and changing the numbering leads to changes in the form of the transition matrix Π. The Proposition 2.2 states that there is an enumeration of vertices that converts Π to the canonical form

$$\Pi = \begin{pmatrix} \Pi_1 \cdots\cdots\cdots\cdots \\ \ddots \\ 0 \ \Pi_k \cdots\cdots \\ \ddots \ \ddots \\ 0 \ 0 \ \Pi_s \end{pmatrix},$$

where each diagonal block Π_k either corresponds to a strongly connected component of H_k or corresponds to a nonrecurrent vertex and consists of a single zero. Under the diagonal blocks there are only zeros.

Our goal is to develop an algorithm that performs this renumbering. The proposed algorithm is a modification of the algorithm described above, in which the order of removing constructed components determines new numbering of the vertices. The renumbering algorithm is as follows.

(1) Let us walk into the depth of a graph G and arrange the vertices in descending order of the exit time $leave[v]$.
(2) Construct transposed graph G^T. Set $no = 0$.
(3) Determine vertex $v \in G$ with the maximum value $leave[v]$.
(4) Check for the existence of an edge of the form $w \to v$, $w \in G$. If such an edge exists, then go to step 5; otherwise, set $H(v) = \{v\}$ and go to step 6.
(5) Start a walk deep into the graph G^T from the vertex v to find the component $H(v)$.
(6) Let the set $H(v)$ contain k vertices, enumerate vertices $H(v)$ in succession from $no + 1$ to $no + k$. Thus, each vertex $u \in H(v)$ is assigned the number $no(u)$. Set $no := no + k$. Remove the vertices $H(v)$ from G and G^T together with the edges outgoing from them. If $G \neq \emptyset$, then go to step 3, otherwise the algorithm is complete.

As a result, each vertex u gets a number $no(u)$ and the resulting numbering converts Π to its canonical form. The algorithm uses memory and time proportional to the number of vertices and edges in the graph.

11.6 Extreme Cycles

Let G be a graph such that each vertex has at least one outgoing edge. Note that this graph always has a closed path, because we can construct an infinitely path forward and the number of vertices is finite. Consequently, the vertices on this path are repeated and give closed paths. Let each edge $e = (i \to j)$ be associated with a real number $c(e) = c_{ij}$. A matrix $A = \left(c_{ij}\right)$ defines the graph's frame. Let $\omega = \left\{i_0, i_1, \cdots, i_p = i_0\right\}$ be a cycle. The average frame over the cycle

$$\eta(\omega) = \sum_{e \in \omega} \frac{c(e)}{p},$$

is called the characteristic of ω. Consider the problem of finding a cycle with maximum or minimum characteristics. Such a cycle is called extreme. In the works [4, 7–10] various methods for solving the problem are discussed. To solve the problem, we consider a potential method. This method was proposed by Romanovsky [10] and is the best.

Romanovsky's Algorithm

Without loss of generality, let us assume that the graph has one strong connectivity component. If this is not the case, then the proposed algorithm must be applied separately to each strong connectivity component. The notation $G = (M, N)$ will denote a directed graph, with a set of vertices M and a set of arcs N. Let the graph G have a framing $A = \left(c_{ij}\right)$. Consider an algorithm for finding a minimum-characteristic cycle. For an arc $j \in N$, $b(j)$ is the starting vertex, and $e(j)$ is the endpoint of that arc j. Let $\omega = \left\{i_0, i_1, \cdots, i_p = i_0\right\}$ be the cycle and z be its characteristic. According to the theory [10, 11], z is the required minimum average if there exists a set of values $v(i), i \in M$ that satisfy the inequalities:

$$z + v(e(j)) - v(b(j)) \leq c(j), j \in N,$$

moreover, the equality

$$z + v(e(j)) - v(b(j)) = c(j), j \in N_\omega$$

is executed on the cycle ω. This system has a unique solution up to an arbitrary constant for variables $v(i)$ (which allows us to refer to them as potentials). We say that a tree has a root i_0 if it contains all paths from i_0 to every other vertex. We consider a tree with one additional arc that generates a (single) cycle ω. To construct the tree, we divide the set of all vertices into three subsets: $M = M_0 \cup M_1 \cup M_2$, where M_0 is the set of vertices that have been processed by the algorithm, M_1 is the set of vertices waiting for processing, and M_2 is the set of vertices that are not reached by the algorithm yet. The vertices M_1 will be ordered. At any given moment, there is tree with the set of vertices $M_0 \cup M_1$ and a root at some vertex i_0. Moreover, for every vertex in $M_0 \cup M_1$ the potential v is calculated (possibly it is not a final result).

The algorithm for constructing the tree begins with some a basic cycle $\omega = \{i_0, i_1, \cdots, i_p = i_0\}$ with vertices and edges (M_ω, N_ω), $M_0 = \emptyset$.

(1) For a given cycle (M_ω, N_ω), compute its characteristic z. Set $v(i_0) = 0$. For all arcs $j \in N_\omega$ calculate the potential

$$z + v(e(j)) - v(b(j)) = c(j)$$

in the order of the vertices on the cycle. Then, form the tree from the cycle by eliminating a single arc incoming to i_0.

(2) Include all vertices of M_ω in M_1 by forming a list.
(3) Act as follows (vertex processing) while the list M_1 is nonempty.

3.1 Eliminate the first vertex i_1 from the list and transform it to M_0.
3.2 For every arc $j = (i_1 \to i_2)$, set the potential

$$w = v(i_1) + c(j) - z.$$

If $i_2 \in M_2$, then move the vertex i_2 from M_2 to M_1 by placing it at the end of the list. Then set $v(i_2) = w$ and join the arc j to the tree.

If $i_2 \in M_0 \cup M_1$, compare w with $v(i_2)$.

In the case of $w \geq v(i_2)$, we complete the arc processing and go to step 3.2.

Otherwise ($w < v(i_2)$), verify whether the vertex i_2 precedes the vertex i_1 in the tree.

(a) If i_2 precedes i_1 then the arc j closes the cycle with characteristic less than z. The new contour is taken for the basic one and the algorithm is repeated, that is, go to step 1.
(b) If i_2 does not precede i_1, set $v(i_2) = w$, include the arc j into the tree by replacing it with some other arc with the endpoint i_2. If i_2 is already located in M_0, transform it to M_1 for repeated processing by placing it at the beginning of the list. Go to step 3.1.
(4) If the computation reaches this stage ($M_1 = \emptyset, M_2 = \emptyset$), this means that the system of potentials satisfying the required inequalities is constructed and the basic cycle has the minimal characteristic.

Choice of initial basic cycle

The described algorithm begins work by selecting a basic contour (cycle), which can be arbitrary. However, the algorithm works faster if the initial contour is close to the desired one. To select the initial basic contour, I.V. Romanovsky proposed a fairly effective procedure [4].

(1) For every vertex $i \in M$ choose an arc $j(i)$ starting in i with the minimal $c(j)$. The set of the selected arcs is denoted by N^*. For each vertex i calculate its degree $s[i]$, that is the number of the above-selected arcs incoming to this vertex.
(2) Form a list of vertices M^* of zero-degree vertices.
(3) While the list is non-empty, proceed as follows.

3.1 Eliminate the first vertex i_1 from the list.
3.2 Eliminate $j(i_1)$ from N^*.
3.3 Decrease the degree of vertex $i_2 = e(j)$ by 1. If the obtained degree is 0, include i_2 into the list M^*.

(4) When the list becomes empty $M^* = \emptyset$, the arcs in N^* form isolated cycles. Among these cycles, select one with minimal characteristic.

References

1. Katok, A., Hasselblatt, B.: Introduction to the Modern Theory of Dynamical Systems. Cambridge University Press (1995)
2. Nemytski, V.V., Stepanov, V.V.: Qualitative Theory of Dynamical Systems. Princeton University Press. Russian original 1949 (1960)
3. Kolmogorov, A.N., Fomin, S.V.: Elements of the Theory of Functions and Functional Analysis. NY (1999)
4. Romanovsky, J.V.: Algorithms for Solving Extremal Problems. Nauka, Moscow (1977). (in Russian)
5. Sharir, M.: A strong-connectivity algorithm and its application to data flow analysis. Comput. Math. Appl. **7**, 67–72 (1981)
6. Siek, J.G., Lee, L.Q., Lumsdaine, A.: C++ Boost Graph Library. Addison-Wesley, Boston—San Francisco—New York (2001)
7. Cochet-Terrason, J., Cohen, G., Gaubert, S., McGettrick, M., Quadrat, J.-P.: Numerical computation of spectral elements in max-plus algebra. IFAC Conference on System Structure and Control (1998)
8. Georgiadis, L., Goldberg, A.V., Tarjan, R.E., Werneck, R.F.: An experimental study of minimum mean cycle algorithms. In: Proceedings of 6th International Workshop on Algorithm Engineering and Experiments, ALENEX, 1–13 (2009)
9. Karp, R.M.: Characterization of the minimum mean-cycle in a digraph. Discret. Math. **23**, 309–311 (1978)
10. Romanovsky, J.V.: Optimization and stationary control of discrete deterministic process in dynamic programming. Kibernetika, 2 (1967), 66–78 (in Russian); Engl. transl.: Cybernetics, 3 (1967).
11. Osipenko, G.S., Romanovsky, J.V., Ampilova, N.B., Petrenko, E.I.: Computation of Morse spectrum. J. Math. Sci. **120**(2), 1155–1166 (2004)

Bibliography

12. Andronov, A.A., Pontryagin, L.S.: Rough systems. In: Reports of the USSR Academy of Sciences **14**(4), 247–250 (in Russian) (1937)

13. Boltzmann, L.: Sitzungber. Kaiserl. Akad. Wiss. **66**(2), 275 (1872)
14. Clausius, R.: Ueber verschiedene fur die Anwendung bequeme Formen der Hauptgleichungen der mechanischen Warmetheorie. Annalen der Physic und Chemie, Bd. **125**(7), 353–400 (1865)
15. Kolmogorov, A.N.: New metrics invariants of transitive dynamical systems and automorphisms of Lebesgue spaces. Doklady AN SSSR **119**(5), 861–864 (1958)
16. Kolmogorov, A.N.: A new metric invariant for transitive dynamical systems and automorphisms of Lebesgue spaces. Topology and ordinary differential equations, dynamical systems. Trudy Mat. Inst. Steklov **169**, 94–98 (1985)
17. Lefschetz, S.: Differential equations: geometric theory. Interscience publishers, NY (1957)
18. Mañé, R.: Characterization of AS diffeomorphisms. Lect. Notes in Math. **597**, 389–394 (1977)
19. Mañé, R.: A proof of the C^1 stability conjecture. Publ. Math. Inst. Higher Studies **66**, 161–610 (1988)
20. Osipenko, G.S.: Localization of the chain recurrent set by symbolic dynamics methods. In: Proceedings of Dynamics Systems and Applications, vol. 1, Dynamic Publishers Inc., pp. 227–282 (1994)
21. Osipenko, G.S.: Spectrum of a dynamical system and applied symbolic dynamics. J. Math. Anal. Appl. **252**(2), 587–616 (2000)
22. Osipenko, G.: Symbolic images and invariant measures of dynamical systems. Ergod. Theory Dyn. Syst. **30**, 1217–1237 (2010)
23. Osipenko, G.S.: Lyapunov exponents and invariant measures on projective bundle. Mathematical Notes **101**(3–4), 666–676 (2017)
24. Osipenko, G.S.: The spectrum of the averaging of a function over pseudotrajectories of a dynamical system. Sbornic: Math. **209**(8), 1211–1233 (2018)
25. Osipenko, G.S.: Mean convergence of periodic pseudotrajectories and invariant measures of dynamical systems. Mathematical Notes **108**(6), 854–866. Pleiades Publishing, Ltd., 2020. Russian text published in Matematicheskie Zametki (2020)
26. Osipenko, G.S.: Encodings of trajectories and invariant measures. Sbornic: Math. **211**(7) 1041–1064 (2020)
27. Robbin, J.: A structural stability theorem, Annas of Mathematics **94**(3), 447–493 (1971)
28. Robinson, C.: Structural stability of C^1-diffeomorphism. J. Differ. Equ. **22**(1), 28–73 (1976)
29. Robinson, C.: Stability theorems and hyperbolicity in dynamical systems. Rocky Mount. J. Math. **7**(3), 425–437 (1977)
30. Sacker, R., Sell, G.: Existence of dichotomies and invariant splittings for linear differential systems I-III. J. Diff. Eq. **15**(3), 429–458 (1974); **22**(2), 476–522 (1976)
31. Salamon, D., Zehnder, E.: Flows on vector bundles and hyperbolic sets. Trans. AMS **306**(2), 623–649 (1988)
32. Shannon, C.: A mathematical theory of communication. Bell Syst. Tech. J. **27**(3), 379–423 (1948)
33. Shannon, C.: A mathematical theory of information. Bell Syst. Tech. J. **27**(4), 623–666 (1948)

MIX
Papier aus verantwortungsvollen Quellen
Paper from responsible sources
FSC® C105338

If you have any concerns about our products,
you can contact us on
ProductSafety@springernature.com

In case Publisher is established outside the EU,
the EU authorized representative is:
Springer Nature Customer Service Center GmbH
Europaplatz 3, 69115 Heidelberg, Germany

Printed by Libri Plureos GmbH
in Hamburg, Germany